THE LOCOMOTI
OF THE
GREAT WESTERN RAILWAY

Part Fourteen

NAMES AND THEIR ORIGINS
RAILMOTOR SERVICES WAR SERVICE
THE COMPLETE PRESERVATION STORY

Published by

THE RAILWAY CORRESPONDENCE AND TRAVEL SOCIETY

1993

THE LOCOMOTIVES OF THE GREAT WESTERN RAILWAY

Part Fourteen

CONTENTS

Photograph] [John Davies

The up "Carmarthen Express", hauled by No. 6000 *King George V,* **climbing Cockett Bank (ruling gradient 1 in 50) on a dull afternoon in September 1987.**

P1

FOREWORD

It is not without a hint of trepidation that a mere Scot, even with Hampshire connections, sets about writing a Foreword to a volume about the Great Western. On reflection, however, having travelled (I am reliably informed) to Redruth on the *Cornish Riviera* pre-war, having crossed Crumlin Viaduct behind No. 3655, having on several occasions "flown" through Little Somerford behind Nos. 7018 or 7019 in their Indian Summer on the up *Bristolian*, having travelled the whole surviving G.W.R. system as at 1964, and having ultimately become the Chief Passenger Manager of what still seemed a very real Great Western Railway indeed, perhaps I can overcome my trepidation. Nevertheless I feel honoured, both in G.W.R. and in R.C.T.S. terms, because this volume is the last of a historic series.

By coincidence, I write these words after paying my last respects to another "last of the line", my good friend Harry Roberts, the last Works Manager at Swindon. His efforts to ward off the closure of the Works have earned him a permanent place of affection in Swindon's heart.

How timely therefore, with the "Western Region" title duly buried and a new era dawning (in which the Great Western name is revived in the InterCity context and more likely to withstand privatisation changes than most other titles) that the Society should now produce the "summing-up" volume to its history of THE LOCOMOTIVES OF THE GREAT WESTERN RAILWAY. This book represents the culmination of over forty years of voluntary effort by a team of dedicated experts who deserve our lasting gratitude. Sadly, most of the original team have passed away, so we must ask Ken Davies, last of the original group and compiler of Part 14 to accept our thanks on behalf of them all. Surely the term "labour of love" can rarely have been more appropriate.

Hugh Gould,
President

August 1992

This drawing of Crumlin Viaduct, reproduced from E. T. MacDermot's *History of the Great Western Railway* **was made very many years before the President's journey referred to above.**

INTRODUCTION

It is over forty years since Part One of the series covering the Locomotives of the Great Western Railway was published, while the last steam-hauled scheduled passenger train left Paddington as long ago as 12th June 1965 (Fig. M133, Part 12).

The demand for information on the subject continues unabated, while interest in Great Western matters has been kept alive by extensive preservation of the steam locomotives, diesel railcars and auto trailers which have been described in the various Parts of the series. Of the 157 steam locomotives which have been rescued from the cutting-up torch, all but a handful survive although it will be many years before some of them run again. It is fitting to pay tribute here to all those devoted to the cause of preservation because by the time that the task is complete, many £millions will have been spent on locomotive restoration alone, with a great deal also spent on preserving whole railways on which to run steam trains and depots to maintain them or to display them. A high proportion of the locomotives can be seen in typical Great Western settings but some have found their way to other preserved railways, with No. 4079 *Pendennis Castle* still providing the pleasure of a ride behind a steam locomotive in Australia.

Parts Two to Eleven inclusive dealt with individual classes and with absorbed railways while Parts Twelve and Thirteen brought the history up to date as at May 1974 and December 1983 respectively. Continued interest in the subject by numerous researchers has produced enough additional information to justify a very comprehensive Part Fourteen and to tie up most (but not all) of the loose ends in previous Parts and this includes much previously unpublished official data. Much more information has come to light about the locomotives sold out of service, even up to a hundred years ago, while research has revealed the necessity for corrections and, happily only in a few cases, the need for "corrections to corrections" but the compilers would rather eat humble pie than perpetuate wrong information. Some readers may feel that "Fourteen" goes into too much detail but as it will be the last Part in the series it has been made as comprehensive as possible.

No history of G.W.R. Locomotives would be complete without reference to the part played by Messrs Woodham Brothers of Barry which resulted in so many examples becoming subjects of actual and proposed restoration and preservation. This has justified a chapter to itself and as the yard at Barry was the sole remaining source of withdrawn locomotives, their disposal has enabled the overall preservation scene to be recorded in an updated chapter, replacing that included in Part Thirteen.

Great Western locomotives have featured prominently in the haulage of special charter trains over British Rail tracks; before they are allowed to do this B.R. requires them to pass stringent tests and it is a tribute to those who restore and maintain their locomotives that they reach the high standards demanded. A separate chapter gives a reasonably complete record of the nature and extent of the workings since Part Thirteen was published, together with attendances at special events.

For many years we have wanted to publish a complete story about the origins and choices of the names given to the locomotives of the G.W.R. and of its absorbed companies but nobody willing to carry out the comprehensive research had been found until we were fortunate to have the help of Bill Peto, an authority on the subject. This accounts for about one half of Part Fourteen and we therefore co-opted him on to the Editorial staff; he has undertaken this daunting task with great enthusiasm. Michael Cook, an authority on industrial locomotives, especially those belonging to contractors, has also been an editor in all but name in so far as the addenda to Parts 3 and 10 are concerned.

Although much of the information has been given in piecemeal fashion in the various Parts, it has been felt appropriate to summarise the changes under British Railways, with particular reference to liveries, especially as the past was happily revived by the use of a quasi-G.W. livery in the late 1950's. Sadly, time has taken its toll and the G.W. story is virtually over. Steam working over the "Great Western" part of the Western Region finished at the end of 1965 (there was a last working on 3/1/66), all but 2-6-2T Nos. 7, 8 & 9 had been withdrawn from other Regions' sheds by November 1966 and the last three 0-6-0PT in London Transport service were officially withdrawn in June 1971 (although No. 7752 as L.T.No. L94 worked a special "last steam train" on 6th July 1971). The last ex-G.W. locomotive to work at an industrial concern was No. 7754 at the National Coal Board's Mountain Ash

PRESERVATION, OLD AND NEW

Photograph] [Reproduced by courtesy of the National Railway Museum, York

Preserved Broad Gauge 4-2-2 *Lord of the Isles* **at Swindon at the turn of the Century.** P2

Photograph] [P. G. Wright

David Woodham unveils the nameplate of No. 7828 *Odney Manor* **at Llwyfan Cerrig Station on 3rd June 1988 in the presence of its owner, Ken Ryder, and of the Gwili Railway Chairman, Keith Mascetti.** P3

BRITISH RAILWAYS' VERY LAST STEAM TRAIN

Photograph] [Martin Porter/The Observer

1'11½" gauge 2-6-2T No. 7 with the inscription "FAREWELL OLD FRIEND" above its smokebox number heads a Devil's Bridge train on 4th November 1988, the last day on which a steam locomotive owned by British Rail (as successors to the Great Western Railway) worked a scheduled passenger train. Fittingly, No. 7 had been built at Swindon for this line in 1923 and it had been appropriately, if belatedly, named *Owain Glyndŵr* **after one of Wales' most famous men in 1956. The railway, with its three steam locomotives and its rolling stock, was sold en bloc in March 1989.** P4

Aberystwyth — Devil's Bridge — Vale of Rheidol Narrow Gauge Steam Railway

British Rail's only narrow gauge railway running through the magnificent scenery of the Rheidol Valley.
For further details please contact Aberystwyth Railway Station. Telephone Aberystwyth 612378.

Miles			Mondays to Fridays Until 4 November			Saturdays Until 29 October			Sundays Until 30 October			Saturdays 10 and 17 December Sundays 11 and 18 December		
0	Aberystwyth	d	1015		1400	1015		1400	1015		1400	1020		1330
11¾	Devil's Bridge	a	1115		1500	1130		1515	1115		1500	1120		1430
0	Devil's Bridge	d	1200		1600	1210		1555	1200		1600	1150		1515
11¾	Aberystwyth	a	1300		1700	1320		1705	1300		1700	1250		1615

For general notes see front of timetable.
Trains will call by request at Llanbadarn, Glanrafon, Capel Bangor, Nantyronen, Aberffrwd, Rheidol Falls and Rhiwfron. Passengers wishing to alight must inform the Guard and those wishing to join must give a clear hand signal to the Driver.
At weekends certain trains may be diesel hauled.

The above is reproduced from page 1524 of the British Rail Passenger Timetable for the period 3rd October 1988 to 14th May 1989. Although trains are shown on Saturdays and Sundays in December, these were operated as "Santa Specials" and were not conventional scheduled services, hence the ceremonial last run on 4th November. The foot of column 2, page M134, Part 12, refers to the last *standard gauge* **steam train.**

NOT WHAT THEY SEEM

Photograph] [P. G. Wright

Robert Stephenson & Hawthorn's No. 7058 of 1942 disguised as G.W. No. 1144 (Hawthorn Leslie's No. 2781 of 1909) at Bronwydd Arms, Gwili Railway in June 1991. P5

Photograph] [K Surman

No. 5051 has its correct number on the buffer beam although temporarily fitted with the name and number plates of No. 5086 *Viscount Horne* **at Didcot Railway Centre on 7th May 1989.** P6

Photograph] [Mick & Joan Roberts

No. 9466 temporarily renumbered 9404 for G.W. 150 Celebrations at Didcot Railway Centre, 23rd May 1985. P7

Photograph] [D. E. White

No. 9629 on the forecourt of Cardiff City Centre's Marriott Hotel on 11th September 1986 has a dummy boiler. P8

PRESERVATION AND REPLICATION

Photograph] [Charles Whetmath

No. 5572, on loan from Didcot Railway Centre to the West Somerset Railway, runs into Blue Anchor station on a Bishops Lydeard train. P9

Photograph] [Mark Wilkins

No. 4588 runs round a Dart Valley train which has arrived from Buckfastleigh at Totnes British Rail station on 21st September 1986. P10

Photograph] [Michael Baker

Iron Duke **and** *King George V* **at Old Oak Common, 15th September 1985.** P11

colliery (Fig. N18, Part 13 shows it running on G.W. metals on 12/6/74 and in July 1974 it was shown as "spare to diesel"). The three ex-Vale of Rheidol 2-6-2T Nos. 7-9, the last steam locomotives of all in B.R. stock, were sold along with the railway to a private company in March 1989, leaving auto trailer No. 233, as "TEST CAR 1", at Derby as the sole survivor of the B.R. stock which has been described in previous Parts. There remains a tenuous link with Swindon Works in the shape of a Swindon-built six-wheeled diesel shunter; the only survivor in B.R. stock of all the steam and diesel locomotives built for British Railways at Swindon, it is used on maintenance trains in the Isle of Wight. Sic transit gloria !

It has been pleasing to receive authoritative contributions to the stories of the Steam Rail Motor cars and of the "Dean Goods" 0-6-0's in both World Wars, which have been extensive enough to justify chapters to themselves. Similarly numerous people have contributed unusual photographs and in choosing illustrations the emphasis has been on reproducing these rather than conventional pictures. Some of them are from tinted picture postcards which do not reproduce as well as a clear black and white print but their value as illustrations (and that of some sub-standard pictures) outweigh these disadvantages.

So many people and organisations have contributed to the series and to Part 14 in particular that any attempt at listing them would inevitably leave some out so this is an expression of the Society's thanks to them all; however, it is appropriate to pay tribute to those who have provided data about the later history of locomotives sold out of main line service and about preserved locomotives and private railways. Bill Peto would like to express his thanks to the countless librarians, archivists and owner/occupiers of stately homes who have contributed so much to his research into the origins of locomotive names.

As to the future, the preservation scene will inevitably change and readers will continue to contribute extra information (and corrections!) to this section and to other aspects of this and earlier Parts. Since "Part 15" et seq are not envisaged, the extra information will be published at intervals in the *Railway Observer*. The compilers of the series have been able to build up an almost complete set of Swindon's "Engine Diagrams" but market research indicates that publishing them as part of the series (Parts 15, 16 and 17) would not be a viable proposition and a set will be deposited in the Society Library. Much detailed information on locomotive allocations from 1901 onwards and on locomotive stock alterations also has been found but, like the "Engine Diagrams", do not justify publication in this series. Details will be lodged in the Library in due course.

Completion and publication of Part Fourteen has been delayed by the untimely passing of D.E. ("Doug") White, who had been the driving force ever since the project was first mooted in the 1940's. Having edited the *Railway Observer* for very many years, Doug knew what to put in and, more importantly, what to leave out and I feel sure that "Fourteen" would have been shorter and more to the point if he had still been around to use his pruning hook.

His picture of No. 9629, see Fig. P8, has been repeated in colour on the flyleaf as a tribute to the one who masterminded the whole project.

Ken Davies, Compiler, Part Fourteen.

A MATTER OF LUCK — THE BARRY SAGA

It is nothing new for considerable numbers of steam locomotives to be massed together awaiting disposal (usually by cutting up) or awaiting the implementation of a refurbishment programme.

The most significant example was at the change of gauge on the Great Western Railway in 1892 which made almost two hundred locomotives redundant. Well over one half of these were convertible to standard gauge and in view of the urgency as many as possible would have been taken into the shops direct from service although most of them and the 84 broad gauge locomotives which could not be converted were assembled in a dump at Swindon. Apart from a couple of small locomotives, all the conversions had been completed by March 1893 and it seems as if the dump of unconvertible locomotives must have been cleared by June 1893 when the two broad gauge "Works Shunters" *Lance* and *Osiris* were cut up (see illustration B108, Part 2).

Following the Grouping in 1923, a number of locomotives from the South Wales railways were gradually assembled at Swindon pending refurbishment and their numbers were much increased late in 1925 when eighty of the R.O.D. 2-8-0's (see page K269, Part 10) were stored until they could be prepared for traffic, an operation which lasted until May 1927. In the meantime, more South Wales locomotives had arrived and were offered for sale, along with some unlikely "bargains" from the G.W.'s own stock of 2-4-0's and 4-4-0's. Most of such sales as were effected were to collieries and similar concerns and one can only assume that the Colonel Stephens railways or their foreign equivalents were the hoped-for source of disposal of the more exotic specimens. Completion of delivery of 0-6-2T No.s 5600-99 and 6600-99 in 1928 had also made at least an equal number of South Wales locomotives redundant and some of these, unfit for sale, added to the assembled company until they could be cut up. Despite much research, no photograph of this large store of locomotives has been found, to compare with Fig. B108 (see above) and Fig. N19 of the Barry Yard in Part 13.

But nothing approached the massacre of Great Western steam locomotives following their replacement by diesel locomotives and railcars under the 1955 Modernisation Plan. On 31st December 1959 there were 14,452 steam locomotives in British Railways' stock, including 2,846 of Great Western origin or design. All but three G.W. locomotives had been withdrawn by November 1966 and every British Rail locomotive except these three (Vale of Rheidol 2-6-2T Nos. 7, 8 and 9) had ceased work by August 1968. The task of disposal of so many locomotives in a short space of time was well beyond the capabilities of the British Railways' works and thus tenders were invited from outside contractors at various points in the British Isles.

One such Company was Messrs Woodham Brothers Ltd. of Barry and this is where a matter of luck emerges. Prior to 1959 the firm was engaged in breaking up B.R. wagons for scrap, at its Barry Docks site, and immediately tendered for locomotives when the opportunity arose in 1959. The Barry location was readily accessible to Great Western locomotives in particular, with rail access to the site which made delivery costs lower than those of sites which were not rail-connected. The yard also was near two docks from which scrap metal could be exported if required. The table below shows the catchment area from which locomotives were acquired, with a preponderance of G.W. examples and only one ex - L.N.E.R. locomotive, B1 Class 4-6-0 No. 61264, which arrived from Nottingham as late as 1968. (This moved later to the Great Central Railway at Loughborough, where it is being restored).

The first locomotives to arrive at Barry Yard were G.W. Nos. 5312/60/92/7 in 1959 followed by another 69 up to the end of 1963 but the cutting-up rate did not keep abreast of arrivals, only 33 having been dealt with in the period. It was easier and quicker to cut up the continual supply of wagons than locomotives!

During the next five years, locomotives of L.M.S.R., S.R. and B.R. origin and the solitary L.N.E.R. example arrived in company with further G.W.R. locomotives, the last of which — Nos. 4920, 6984/90, 7802/12/9/20/1/7/8 and 7927 — reached Barry in 1966. The final total was 288, made up as follows:

G.W.R.	163	of which	65	were cut up
L.M.S.R.	37	" "	2	" " "
S.R.	45	" "	4	" " "
B.R.	42	" "	4	" " "
L.N.E.R.	1	" "	—	" " "
Total	288	" "	75	" " "

Thus 213 locomotives (98 of G.W. design) were saved from cutting-up. One of the S.R. examples, No. 35029, was cut in half longitudinally for display at the National Railway Museum, York, and the other 212 were available for preservation, either in their own right or, in a few cases, as a source of components for other restored locomotives. The highest number of locomotives at Barry at any one time was 221 in 1968, with a preponderance of Great Western examples.

The earlier arrivals at Barry were all cut up and it was September 1968 before the first locomotive left Barry for preservation — L.M.S. Class 4F 0-6-0 No. 43924, destined for the Keighley & Worth Valley line; the first G.W. locomotive to escape the cutter's torch was 2-6-0 No. 5322 which left for Caerphilly in March 1969 where it was restored and first steamed in December 1970. Thereafter there was a steady stream of departures for preservation but it was not until 9th November 1989 when, with one exception, the Barry yard was emptied with the departure of 2-8-0 No. 3845 for Brighton Railway Museum. The exception was 2-6-2T No. 5553 which Woodham Bros. had presented to the Barry Development Partnership for static display in Barry but later exchanged for No. 5538 on the Dean Forest Railway.

The 163 G.W.R. locomotives bought by Woodham's comprised the following classes:

4-6-0	"King"	2	Brought forward		94
"	"Castle"	5	2-6-2T	"5101"	14
"	"Hall"	17	"	"6100"	1
"	"Manor"	8	0-6-2T	"5600"	12
2-8-0	"2800"	16	0-6-0T	"1366"	2
2-6-0	"4300"	10	"	"5400"	3
2-8-2T	"7200"	4	"	"5700"	9
2-8-0T	"4200"	8	"	"6400"	1
2-6-2T	"3150"	1	"	"8750"	9
"	"4500"	23	"	"9400"	18
Carried forward		94	Total		163

Whilst a worthy number of classes of G.W. locomotives have been rescued and many restored, either to working order or cosmetically, it is a matter of regret that neither the "3150" Class 2-6-2T No. 3170 nor any of the three "5400" Class 0-6-0T's survived the cutter's torch at Barry as these classes are not represented in the preservation scene. Nor are the "Grange" 4-6-0 or the "7400" 0-6-0T classes but this cannot be blamed on Woodham Bros. as they did not purchase any examples of these classes.

There was of course mass cutting-up of locomotives at yards belonging to other firms throughout the country but, with two exceptions, none of them deferred cutting-up their stock, much reducing the opportunities for preservation. The exceptions were Messrs Draper of Hull, from whose yard L.M.S. "Black Five" No. 45305 was preserved, and Messrs R. S. Hayes, Bridgend, who used G. W. No. 9642 to shunt their yard for about three years in 1965-67, after which it was sold to Messrs W. T. and D. R. L. Jones for preservation. No. 6430 is in a different category, as described on page P6. British Rail sold several locomotives in running order and preserved a few themselves; there also were a few preserved locomotives of an earlier vintage and details of the scene as a whole are given in the up-dated "Preservation" chapter on page P6. (The foregoing account is based upon "The Barry Locomotive Phenomenon" by Francis Blake and Peter Nicholson (OPC/Hayes Publishing Group 1987) and the information is reproduced with the authors' consent).

PRESERVATION

In common with other main line railways, the Great Western did very little to preserve its locomotive heritage. *North Star* was preserved at Swindon in its as-withdrawn condition from 1871 to 1906 but then broken up "due to shortage of space". The present *North Star* is a full size model of the original locomotive of 1837 which contains some of the original parts, charitably categorising it as a "preserved locomotive" (see Part 2, figs B9 and B10). *Lord of the Isles* similarly was preserved at Swindon from 1884 to 1906 (fig P2) and was broken up for the same reason but in this case the driving wheels, the complete reversing gear, and part of the motion (Pages B20 and N36) were retained and are displayed at the G.W.R. Museum, Swindon. *Tiny*, the smallest broad gauge locomotive, was more fortunate; on withdrawal in 1883 it was put to stationary use and restored as a static locomotive exhibit in 1927. It is now in the Dart Valley Railway museum at Buckfastleigh.

Apart from the frames, wheels and motion of No. 252 (see page C26 and Page P84 of this Part) there were no further attempts at preservation until *City of Truro* was sent to York Railway Museum on withdrawal in March 1931 (Page G37). As recorded on page M92 it was restored to running order, in approximate 1904 condition and with its original running number 3440. Its subsequent exploits justify a separate chapter, later.

The G.W.R. acquired the Wantage Tramway 0-4-0WT *Shannon*, built by George England in 1857, upon closure of the tramway at the end of 1945. Although it must surely have worked into G.W.R. sidings at Wantage Road much earlier, it was not registered to do so until 1941 (No. 209, see Part 13, page N34). Whether this tenuous link with the G.W.R. added to its age in deciding to preserve it is not known but it was sent to Swindon, overhauled and repainted in the spring of 1946, appearing on Wantage Road station as a static exhibit some two years later. It was removed to local storage as from 1/1/66 and was delivered to Didcot Railway Centre on 18/1/69, where it was made fully serviceable by March 1970.

Swindon presented No. 4073 *Caerphilly Castle* to the Science Museum, Kensington, in June 1961 and Nos. 2516, 4003 *Lode Star* and 9400, along with the full-size model of *North Star* to Swindon Museum when it opened in 1962. Nos. 1442 and 2818 were sold as static exhibits and, in conjunction with *Tiny* (see above) and Nos. 921, 1378, 5538 and 9629 all have the "SE" status in the Table. Most of these could probably be set to work again after a modest amount of refurbishment but *North Star* and Nos. 5538 and 9629 could not be made serviceable at reasonable expense. The last two have dummy boilers and that formerly on No. 9629 is to be put on No. 3650 at Didcot. No. 9629 was, exceptionally, bought from Barry by Commonwealth Holiday Inns of Canada and sent to Steamtown, Carnforth, in May 1981 for restoration as a static exhibit. It reached Cardiff in time for the G.W.R. 150th Anniversary Celebrations on 6-10th July 1985 and had arrived at its ultimate destination, a length of track outside Cardiff's Holiday Inn, by April 1986. (Fig. P8).

Towards the end of the steam era, Western Region sold Nos. 3, 4, 822/3, 1338/63/9, 1420/50/66, 1638, 3205, 3217 (as 9017), 4079, 4555, 6000, 6106, 6412/35, 6697, 6998, 7029 and 7808, all more or less in running order. Many of them were delivered to their new owners in steam and are shown in the status column as "RO", reflecting their condition when acquired, although some of them, mainly the smaller locomotives, have been laid aside. The larger locomotives have had (or are having at the time of writing) major overhauls but the situation is always fluid.

Nos. 5764/75/86, 7715/52 were acquired from London Transport more or less in working order but No. 7760 needed an extensive overhaul.

Nos. 12, 426/50, 813, 1340, 1501, 3650, 6430, 7714/54, 9600/42, B.P. & G.V. 2 *Pontyberem* and P.&M. 2 came from a variety of sources as shown in the Table or in earlier Parts. Of these, Nos. 12, 450 (as Taff Vale No. 28 — see fig P36) and 9642 are serviceable; P.&M. 2 has been scrapped and no work is being done on *Pontyberem* but the rest are in various stages of restoration with No. 426 planned to appear as Taff Vale No. 85. No. 813 has been in steam on several occasions but has been dogged by teething troubles and is not yet completely serviceable.

When condemned locomotives were sold to scrap merchants, there was a ban on resale to preservationists but this seems to have been overcome in the case of No. 6430 (sold to J. Cashmore Ltd, see Part 12, page M88) by describing it as a source of spare parts for Nos. 6412/35 which also were bought by the Dart Valley Railway.

No. 9642, sold to R. S. Hayes (see page M85) had evidently been considered to have worked its passage by shunting their scrapyard for almost three years.

This leaves by far the largest source of former G.W. locomotives as Messrs Woodham Bros. yard at Barry described in detail in the chapter headed A MATTER OF LUCK. These are 98 locomotives of G.W. design and Class 8F 2-8-0 No. 8431 which justifies inclusion in the G.W. list because it was built at Swindon. In the early days, some of the locomotives at Barry were fit to be moved by rail but none was anywhere near running order and many of the later sales lacked vital parts. It would be more logical to regard many ex-Barry locomotives as rescued rather than preserved and tribute is due to those people who have taken these locomotives into their care in the knowledge that it will cost more than £100,000 (sometimes a great deal more) to fully restore them. This supersedes the "Woodham" references on pages N3 and N6.

Four of these locomotives were acquired as sources of spare components for restoration of other locomotives. Nos. 2873 and 5043 at Tyseley come into this category, No. 3612 on the Severn Valley Railway was utilised in this way and the remains cut up in September 1990 while No. 5532 which the Dean Forest Railway bought for spares has provided a set of frames which are to be combined with the boiler and wheels of No. 5538 and will run as No. 5532. To further complicate the matter, No. 5553 which had been retained at Woodham's yard with a view to becoming a static exhibit in Barry Town Centre was found to be in better condition than the Llangollen Railway's No. 5538 and the latter's frames moved via the Dean Forest Railway (where it acquired the tanks and wheels off their No. 5541), to Steamtown, Carnforth. "5538+5541" was restored as a static exhibit, including a dummy boiler, and the locomotive is more appropriately displayed on the promenade at Barry Island, close to the site of some of the most intensive provincial steam passenger workings.

The Vale of Rheidol 2-6-2T Nos. 7, 8 and 9 continued in British Rail service until the railway, along with its locomotives (in working order) was sold to the Brecon Mountain Railway Company with effect from 31st March 1989. (Fig. P4).

There are no more candidates for preservation and the final situation can be summarised as under. *North Star* perhaps should be included with *Iron Duke* and *Fire Fly* as it is not truly preserved but it has been around for a long time:

Preserved by G.W.R. and B.R.W.R. as static exhibits	6
Preserved by G.W.R. and B.R.W.R. in running order (No. 3440)	1
Preserved by G.W.R. but now in running order (*Shannon*)	1
Other static exhibits (total 6 less 2 included under Woodham)	4
Sold in more or less running order by B.R.W.R.	23
Vale of Rheidol locomotives sold by B.R.W.R.	3
Acquired from London Transport	6
Acquired from collieries and other miscellaneous sources	14
Obtained from Woodham Bros., Barry (includes 5538 & 9629 "S.E.")	99
Total	157

Table 1 on pages N7 to N14 of Part 13 is replaced by the new Table 1 but for completeness, actual errors in Part 13 are set out here. The main discrepancy is that anticipated completion dates, given by owners, did not transpire for Nos. 2807, 3822, 4110/44, 4920, 5193, 5637, 5900/72, 6024, 6634/95, 6989/90, 7200, 7754, 7802/21/2/8, 7903, 9466 and Auto Trailer No. 38. No. 9681 was restored, in 9/84, and Nos. 3822, 4920, 5900, 6024, 6990, 7802/22/8, 9466 and Trailer 38 were restored at a later date, see new Table 1. The status "RO" against Nos. 5322 and 6960 was that which applied when they reached the "Current Location" shown on pages N10/11. The actual dates of restoration appear in new Table 1. Other corrections to be noted are: No. 1340 was not restored in 1983; the correct location of No. 1378 is Scolton Manor Museum; No. 1501 arrived on the S.V.R. 10/70; No. 5541 arrived on the D.F.R. in 10/72; No. 5643 delete Date of Arrival 9/71, see note (33) to new Table 1; No. 5786 reached Bulmer's, Hereford in 7/70; No. 6106 arrived at Didcot 11/67; No. 6990 arrived on the G.C.R. in 11/75; No. 7760 was restored in 3/88, not -/83; No. 7827 arrived on the D.V.R., Paignton, in 3/73; L.M.S. No. 8431 was restored in 12/75, not 12/74; No. 9400 moved to Swindon Museum in 4/62; Auto Trailer No. 92 arrived Swindon 6/70, not 4/70 as shown on page N15; No. 190 was not restored in 1970 and remains unrestored; No. 231 arrived Didcot 12/67.

The status of the 157 locomotivies at the time of writing is given below (the total of 156 given in line 40 column 1, page N3 excluded P. & M. No. 2, which had by then been scrapped, see page N39):

In running order when first acquired, status "RO"	28
Restored to running order by date shown as "Rxx/XX"	42
Actively under restoration, status "UR"	47
Awaiting start of restoration apart from cosmetic treatment, status "AR"	24
Static exhibits, status "SE"	12
Acquired as a source of spares, status "FS"	2
Scrapped: No. 3612 after removing usable parts; P.&.M. No. 2	2
Total	157

The information given in the "Current Location" column is the best available at the time of writing. Of those "in works", the future location of Nos. 4248, 4612 and 5553 has not been decided. The same applies to No. 6000, see later.

Having preserved or restored Great Western locomotives, the obvious development is to run them on what was the Great Western Railway. Only a few have been passed by British Rail for main line running and their exploits are described in another chapter. The remainder are confined, where available, to closed G.W. lines and it is not an economic proposition to run over any part of B.R. metals to gain access to a B.R. station (as the Dart Valley Railway found to its cost at Totnes), because safety considerations involve the provision of both a B.R. pilotman driver and guard for the journey over B.R. metals, which in effect means four men doing the work of two. This does not prevent B.R. locomotives working trains on to private lines, to be replaced by private locomotives (as happens on the West Somerset and Severn Valley Railways) or to arrange cross-platform interchange, which is possible at Bodmin Parkway station (formerly Bodmin Road), whence the Bodmin & Wenford Railway plc operates steam trains to Bodmin General with track in place to extend to Boscarne Junction later. Unique among ex-G.W.R. preserved railways, the B.&W. ran a freight service from Bodmin connecting with the private successor to the B.R. "Speedlink" service, which carried electrical goods to North of England destinations until December 1992. The remaining 7.5 miles of this 13.5 mile route was part of the L.&S.W.R., but as it served china clay workings there is a distinct possibility that traffic of this kind also could be developed as there is similar traffic from other places in Cornwall.

Taking other railways based on closed lines in alphabetical order (and with only a brief reference to those described in Part 13, pages N4 to N6), the Cholsey & Wallingford Railway has possession of this branch except for about 1/3rd mile at the Wallingford end and hopes before long to restore track into the former bay platform at Cholsey which will afford cross-platform connections with B.R. trains.

The newest recruit to the preservation movement is the Chinnor & Princes Risborough Railway, located on part of the Watlington & Princes Risborough branch. The company seeks a Light Railway Order for a mile and a half of the branch. There are numerous cases of preserved locomotives being loaned to other operational companies but here Buckinghamshire Railway Centre has allowed the use of their "Sentinel" locomotive, G.W. No. 12, on work connected with re-opening the line.

The Dart Valley Light Railway plc (page N4) has leased the Totnes-Buckfastleigh branch, now known as the Primrose Line, to the South Devon Railway Trust but continues to run the Paignton & Dartmouth Steam Railway. Since locomotives can appear on both lines at different times, they are shown for brevity only in the Table as "Dart Valley Railway". Trains on the Primrose Line now terminate at a station known as Littlehempston Riverside on the far side of the River Dart from the B.R. station and it will be necessary to revive the footbridge project (Page N4).

The Dean Forest Railway extended its operations to a new station known as Lakeside on 18th August 1991 which involved re-opening an existing level crossing over the A48 Trunk Road at Lydney. This is probably the first example of a private railway running trains across a major road, but the right of way would have been established by long usage from the days of the Severn & Wye Railway. The next stage will take the railway to a station near the B.R. Lydney Station, making spot connections with B.R. trains possible. Later still the trains will be extended northwards to Parkend, making a total track length of about four miles (see page N6).

The East Somerset Railway operates passenger trains from a restored station at Cranmore for two miles in the Shepton Mallet direction. Access to the B.R. main line can be obtained via the Merehead Quarry connection (page N5).

The ambitious Gloucestershire & Warwickshire Railway (page N5) has now reached Gretton, four miles from Toddington and is continuing towards

Cheltenham, its first major objective. When this is achieved the Company intends to extend to Honeybourne, where a connection could be made with British Rail trains, and ultimately to Stratford-upon-Avon, 29 miles from Cheltenham.

The Gwili Railway (page N5) is extending its track to serve Conwil station but still has no G.W. locomotives of its own although it has hired more of them in proportion to its length than any other preserved railway! (The 0-4-0ST masquerading as G.W. No. 1144 in the 1991 season (Fig. P5) is actually an almost identical but much later example of the Hawthorn Leslie traditional design, Robert Stephenson & Hawthorn's No. 7058/1942). A compliment was paid to David ("Dai") Woodham of Woodham Bros. (see "The Barry Saga"), when restored No. 7828 "called" at the Barry yard en route to the Gwili Railway. "Dai" was the guest of honour on the footplate on its first service run on 3rd June 1988 (Fig. P3).

The Llangollen Railway (page N5) is making good progress, now having nearly five miles of track in situ with a planned extension to Carrog (7.25 miles) and the intention of reaching Corwen (9.75 miles) as soon as possible. There are no firm plans for the obvious eastern extension from Llangollen to the former "Barmouth Bay" at Ruabon B.R. station but there is informal support for the idea, which would result in a 16-mile long preserved railway.

The Plym Valley Railway is based on the former G.W. station at Marsh Mills on the Plymouth-Tavistock-Launceston branch and has 1.5 miles of track in place with eventual extension to Bickleigh, a further 2.5 miles away. Nos. 3802 and 7229 came here from Barry but have since moved elsewhere for restoration. A Swindon-built B.R. 4-6-0 No. 75079 is here, under a protective coat of paint.

The Pontypool and Blaenavon Railway will be unable to live up to its name as road construction has blocked access to Pontypool and the extent of the system will be from Abersychan to Waenavon, almost entirely along a former L.M.S. (L.&N.W.R.) branch, and about six miles long, with two miles of track in place. Five G.W. locomotives rescued from Woodham's yard at Barry await restoration.

The main addition to the account of the Severn Valley Railway on page N4 is to record the opening of the Company's own station near Kidderminster B.R. station.

The Swansea Vale Railway is another borderline case. The original railway of this name became part of the Midland Railway and then of the L.M.S. but made end-on connections with the G.W.R. main line at Swansea Valley Junction, with the Swansea Docks railway system, and with the Neath & Brecon at Ynysygeinon Junction, thus effectively land-locking it. In addition, it logically became part of B.R.'s Western Region and was worked mainly by "6700" class 0-6-0PT in its latter days. It is currently the home of two G.W. locomotives.

Most preserved railways have to work very hard for their living and the Swindon & Cricklade (page N5) is one of the luckier ones in having Local Authority assistance in constructing a diversion to a new station to be built by B.R., some two miles out of Swindon on the Gloucester line. This will afford cross-platform connections. Without going into too much detail, the Local Authority found it cheaper to build an embankment for the railway than to cart a very large quantity of excavated material to a more distant tip.

The Talyllyn Railway was not part of the G.W.R. but merits inclusion here as it set the scene for locomotive preservation in acquiring G.W. Nos. 3 and 4 as long ago as 1951. The Tywyn terminus is not far from the B.R. station.

The Telford Steam Railway, operated by the Telford-Horsehay Steam Trust, has advanced plans to restore 1.5 miles of the former Wellington-Craven Arms line from Horsehay & Dawley station to a point near Lightmoor station. Steam trains currently operate over half a mile of a former siding and use a former industrial locomotive shed at Horsehay.

The Brecon Mountain Railway Company advise that they intend to continue to operate the Vale of Rheidol Railway as before, the locomotives retaining their numbers 7, 8 & 9. (Page N5). The stock is being converted to air-braking and No. 9 has been modified to suit. The Welshpool and Llanfair Railway (of 2' 6" gauge and not as shown on page N5) continues as before; No. 822 has been on tour of various railway sites in the past few years but there are plans to reinstate it as a working locomotive.

The locomotive scene on the West Somerset Railway has changed radically from that described on page N5 and the line now has a "Great Western" flavour with Nos. 3205, 4561 and 6412 operational and Nos. 3850, 4160, 5542 and 7820 all well in hand. The independent track into Taunton referred to on page N5 has been lifted but there is a physical connection with B.R. through the Taunton Cider Co's siding at Norton Fitzwarren.

The Dowlais Cae Harris scheme referred to at the foot of column 2, page N5, has fallen through and No. 5668 has gone to Blaenavon instead, see above.

As mentioned in column 1, page N4, Great Western locomotives are also to be found at locations far removed from the Great Western system and there are further additions to this list including: Brighton Museum; East Lancashire Railway, Bury; Fleetwood Locomotive Centre; Herring Bros., Bicester; Kent and East Sussex Railway; Lakeside & Haverthwaite Railway; and Northampton Steam Railway, Chapel Brampton. Steamport, Southport is now known as Southport Steam Centre. No. 5643 has left Steamtown Railway Centre, Carnforth, which now has no G.W. locomotives but has restored No. 5538 for static display on Barry Island promenade. No. 5972 is still at Wakefield but the name of the Company has changed from Procor to Bombardier Prorail.

Most of the above-mentioned companies operate, or should in due course operate, timetabled train services but there is another, important group of preservation sites, often based upon a former Great Western engine shed, where there is space only for a demonstration line. Some of these sites house a large number of preserved locomotives, usually restored "in-house" and there have been frequent cases of hiring to operating railways, not to mention use of their larger locomotives on charter trains on B.R. metals. The hiring situation changes frequently and is described as it happens in the *Railway Observer*; the main line exploits are covered in a separate chapter. (Fig. P9 shows No. 5572 on hire to the West Somerset Railway).

In alphabetical order, Birmingham Railway Museum sponsors "The Steam Depot" in part of the former G. W. Tyseley shed (page N5); "Buckinghamshire Railway Centre" is the new title of the depot at Quainton Road station, described on page N6. It originally was a Metropolitan Railway station and the Centre has a very good relationship with London Underground. 0-6-0PT No. 7715, preserved here, had run as London Transport No. L99 for almost seven years in L.T. service and Quainton stock also includes Metropolitan 0-4-4T No. 1. No. 7715 reached Quainton Road in January 1970 as "L99" but later that year was repainted in B.R. black with brass "7715" number plates. Repainted green in 1981 it ran thus until October 1988 when laid aside for overhaul and at the time of writing is being restored as "L99" in maroon livery again. It will be brought up to B.R. standards for working special trains on the main line; Bulmers' Railway Centre, Hereford (page N6) is technically on a former L.M.S. line, which joins the former G.W. system near Hereford Station; the former B.R. Caerphilly Workshops have been used to restore No. 5322 and to overhaul and repaint 0-6-2T No. 450 in its original guise as Taff Vale No. 28 (fig. P36); Didcot Railway Centre has been described on page N4 and continues to expand its facilities, of which perhaps the most interesting is a length of broad gauge track using original material discovered in South Devon (fig. P12); on a more modest scale is the award-winning restored station at Rowden Mill on the former Worcester-Leominster line, which was to have housed 2-6-2T No. 5553 but at present is the home of a fully-restored Swindon-built diesel shunter No. D2371; Southall Railway Centre, the former G.W. shed, is the home of the G.W.R. Preservation Group and also services main line steam locomotives from time to time; Swindon Museum (page N4) is now known as "G.W.R. Museum, Swindon". The museum's locomotive collection remains intact except that No. 4003 has just gone on loan to York, its place being taken temporarily by No. 6000 pending a decision on the latter's future. There are over 30 nameplates, a set of driving wheels, reversing gear and motion from *Lord of the Isles* and a large photograph collection. There is also an exhibition devoted to the growth of Swindon, its Works and associated social facilities; finally, the Welsh Industrial & Maritime Museum sponsors the Wales Railway Centre where six cosmetically-restored G.W. locomotives are housed at B.R.'s Bute Road Station, Cardiff Docks, pending establishment of a permanent Centre within the Cardiff Bay area. Incidentally, the Museum houses a full-size working model of Trevithick's Pen-y-Darren locomotive of 1804 (and which is steamed occasionally) which is where it all started. But for Trevithick, there may have been no LOCOMOTIVES OF THE GREAT WESTERN RAILWAY to write about!

Some of the larger Centres carry out work on contract for other railways and the most interesting — and appropriate — venture in this field is "Swindon Workshops Limited" which has been set up in the former "R" Machine Shop and part of the former "B" Locomotive Repair Shop, thus maintaining a tradition (see Pages M16/7, Part 12) which started in 1843. There are facilities for working on a number of locomotives at a time, with machine tool capacity to match.

Special mention must be made of the Cambrian Railways Society, a small organisation who nevertheless restored No. 7822 for service on the Llangollen Railway, leaving them only with Auto Trailer No. 163 to represent G.W. stock. This may also be the place to mention the sole survivor among the subjects of this series to remain in British Rail stock; this is Auto Trailer No. 233, in use at British Rail's Derby Technical Centre as "Test Car No. 1". It has been fitted with special bogies and the original have gone to Didcot to replace those under their Trailer No. 231.

On the museum front, the only other change to the situation described in Part 13 is that the Bleadon & Uphill collection has been dispersed and the site closed. The only G.W. example was 0-4-0ST No. 1338, which went to Didcot in March 1987. It is convenient to mention here that No. 1378 at Scolton Manor Museum is the oldest preserved G.W. locomotive, having been built in 1878. The former Wantage Tramway's *Shannon* is 21 years older but it never was included in Great Western stock.

In view of their Swindon origins, it seems appropriate to refer briefly to the preservation of steam and diesel locomotives of non-G.W. design built at Swindon in B.R. days. The steam locomotives are Nos. 46512/21,75014/27/9/69/78/9, 92203/7/12/14/19/20. Diesel hydraulic locomotives are Nos. D821/32, D1010/3/5/23, D9500/2/4/13/6/8/20/1/3-6/8/9/31/7/9/51/3/5, with diesel shunters Nos. D2022-4/41, D2117/9/20/38/48/52/62/78/82/4/92/9, D2371/81. D2179 survives in B.R. stock as No. 03179, currently used as a service locomotive in the Isle of Wight. Together with Auto Trailer No. 233, referred to above, these are the only survivors of the products of Swindon Works to remain in British Railways stock. So far as preservation is concerned, it is fitting that both the last steam (92220) and the last diesel locomotive (D9555) built at Swindon have been kept for posterity.

To prevent confusion, it is necessary to record that 0-6-0PT No. 9466 carried "9404" number plates at Didcot in May and September 1985 (but ran as "9466" in June) for the Anniversary celebrations; this was done at the request of the organisers because the boiler on No. 9466 had originally been on No. 9404. Also, No. 9466 is painted in G.W. green, which applied only to Nos. 9400-9; the other 200 members of the class were painted unlined black. (Fig. P7). This picture appears under the heading "NOT WHAT THEY SEEM" and has a logical explanation but the short-term numbering of an 0-4-0ST with no G.W. connection as "1144" (Fig. P5) and the "renumbering" of the preserved "Sentinel", G.W. No. 12, as L.N.E.R. No. 49 are illogical. Fixing of cherished name and number plates to preserved locomotives for photographic purposes is no doubt a source of pleasure to their owners but could be confusing later on. An example is No. 5051 posing as No. 5086 (Fig. P6) and Nos. 5900, 6998 and 7808 at Didcot (and maybe named locomotives at other preservation sites) also have had other plates affixed. On the other hand, "5700" Class 0-6-0PT which had London Transport numbers late in their careers can, and do, logically appear under either number and in either livery.

Since the account in Column 2, page N7 was written, *Iron Duke* has been completed. Its complicated history started with donation by the National Coal Board of two standard "Austerity" 0-6-0ST to the Science Museum in 1981. These were Robert Stephenson & Hawthorn's No. 7135 of 1944, W.D. No. 75195, later N.C.B. *Gwyneth* at Gresford Colliery near Wrexham and at Bickershaw, Lancs; and Hunslet Engine Co's No. 3696 of 1950, N.C.B. *Respite*, used at various Lancashire collieries. Both were sent to Resco (Railways) Ltd at Woolwich, London, with the intention of incorporating usable parts in *Iron Duke*. Neither was complete and *Gwyneth* was dismantled to provide the boiler, some of the boiler fittings and valves, the cylinder block and the motion in its entirety, except for a small number of parts from *Respite*. The latter was not dismantled and went to the National Railway Museum in February 1985. *Iron Duke* had been completed in 1984 and, joined by a tender and two replica broad gauge carriages, all built in British Rail's Cardiff Workshops, took part in a formal naming ceremony in London's Hyde Park on 3rd April 1985. The Duke of Wellington, a descendant of the "Iron Duke" from which the locomotive took its name, performed the honours; the train moved in steam on a length of broad gauge track. Normally kept at the National Railway Museum, York, it was in steam on a special length of broad gauge track at Old Oak Common on 15/9/85 and, better still, on genuine broad gauge rail (recovered from Burlescombe, Devon) at Didcot (Figs. P11 and P12).

The Fire Fly Trust was set up in 1982 to create a replica of the broad gauge 2-2-2

Fire Fly of 1840 (Part 2, page B13). Started in Bristol, the project was transferred to Didcot Railway Centre in September 1988 and the chassis is ready to receive a boiler.

Amending footnote (j), page N14, the project for converting No. 4942 *Maindy Hall* into a "Saint" Class 4-6-0 seems unlikely to proceed.

Finally, it is re-iterated that 4-2-2 No. 3041 described on pages N6/7, Part 13 is no more than a full-size model and is in no sense a preserved locomotive.

Tailpiece: Two Swindon-built diesel locomotives, Nos. 03128/34 are reported working on a preserved railway in Maldegem, Belgium and Nos. 9515/48/9 seen in 1990 at Madrid's Charmartin shed were said to be stored pending preservation.

TABLE 1
GREAT WESTERN STEAM LOCOMOTIVES

No.	*Present Name*	*Type*	*Page Ref.*	*Current Location*	*Date of Arrival*	*Status*
3	*Sir Haydn**	0-4-2ST	K264 M132	Talyllyn Railway, Tywyn	3/51	R7/51
4	*Edward Thomas**	0-4-2ST	K265	Talyllyn Railway, Tywyn	3/51	R5/52
7	*Owain Glyndŵr**	2-6-2T	K78 M130	Vale of Rheidol Light Railway, Aberystwyth	—	RO
8	*Llywelyn**	2-6-2T	K78 M130	Vale of Rheidol Light Railway, Aberystwyth	See text "	"
9	*Prince of Wales**	2-6-2T	K78 M130	Vale of Rheidol Light Railway, Aberystwyth	"	"
12		4-wheel Sentinel	F8	Buckinghamshire Railway Centre, Aylesbury	5/72	R8/79
426		0-6-2T	K183 M118 M132	Keighley & Worth Valley Railway (1)	1/71	UR
450		0-6-2T	K179 M118	Caerphilly (National Museum of Wales) (2)	6/67	R5/83
813		0-6-0ST	K247 M119	Severn Valley Railway	11/67	UR
822	*The Earl*	0-6-0T	K80 M132	Welshpool & Llanfair Light Railway	7/61	RO
823	*The Countess**	0-6-0T	K80 M117 M132	Welshpool & Llanfair Light Railway	10/62	RO
921		0-4-0ST	K261 M119	Snibston Discovery Park Coalville, Leicestershire	6/67	SE
1338		0-4-0ST	K92	Didcot Railway Centre (3)	3/87	UR
1340	*Trojan*	0-4-0ST	K16 M116	Didcot Railway Centre	3/68	UR

No.	Present Name	Type	Page Ref.	Current Location	Date of Arrival	Status
1363		0-6-0ST	E74 M77	Didcot Railway Centre (4)	10/77	R5/70
1369		0-6-0PT	E74	Dart Valley Railway	4/66	RO
1378	*Margaret*	0-6-0ST	C89 K220 M130	Scolton Manor Museum, Haverfordwest, Dyfed	6/74	SE
1420	*Bulliver**	0-4-2T	F22	Dart Valley Railway	10/65	RO
1442		0-4-2T	F22	Tiverton Museum	10/65	SE
1450	*Ashburton**	0-4-2T	F22	Sold by Dart Valley Railway but new site not known (note 5)		RO
1466		0-4-2T	F22	Didcot Railway Centre (6)	12/67	RO
1501		0-6-0PT	E86 M89 P81	Severn Valley Railway	10/70	UR
1638	*(Dartington)* *	0-6-0PT	E85	Kent and East Sussex Railway (7)	7/92	RO
2180	*Tiny*	0-4-0T	B40 M130	Dart Valley Railway Museum, Buckfastleigh (8)	4/80	SE
2516		0-6-0	D69 M74	Great Western Railway Museum, Swindon	6/62	SE
2807		2-8-0	J19 M111	Gloucestershire & Warwickshire Railway, Toddington	6/81	UR**
2818		2-8-0	J19 M111 M130	National Railway Museum, York (9)	9/75	SE
2857		2-8-0	J19 M111	Severn Valley Railway	8/75	R9/79
2859		2-8-0	J19 M111	Llangollen Railway	11/87	AR
2861		2-8-0	J19 M111	Wales Railway Centre, Cardiff	2/88	AR
2873		2-8-0	J19 M111	Birmingham Railway Museum (10)	4/88	FS
2874		2-8-0	J19 M111	Pontypool & Blaenavon Railway	8/87	AR
2885		2-8-0	J19 M111	Southall Railway Centre	3/81	AR
3205		0-6-0	D81 M75	West Somerset Railway (11)	3/87	RO
3217	*Earl of Berkeley**	4-4-0	G17 M91	Bluebell Railway, Sheffield Park (12)	2/62	RO

No.	Present Name	Type	Page Ref.	Current Location	Date of Arrival	Status
3440	*City of Truro*	4-4-0	G36 M92 P24	Based on National Railway Museum, York, but see text for numerous workings	7/89	RO
3612		0-6-0PT	E77 M83	Severn Valley Railway, for spares	12/78	cut up 9/90
3650		0-6-0PT	E77 M85	Didcot Railway Centre	5/71	UR
3738		0-6-0PT	E77 M84	Didcot Railway Centre	4/74	R4/77
3802		2-8-0	J19 M111	Bodmin & Wenford Railway (13)	2/90	UR
3803		2-8-0	J19 M111	Dart Valley Railway	11/83	UR**
3814		2-8-0	J19 M111	North Yorkshire Moors Railway	7/86	UR
3822		2-8-0	J19 M111	Didcot Railway Centre	5/76	R7/85
3845		2-8-0	J19 M111	Brighton Railway Museum	11/89	AR
3850		2-8-0	J19 M111	West Somerset Railway	3/84	UR
3855		2-8-0	J19 M111	Pontypool & Blaenavon Railway	8/87	AR
3862		2-8-0	J19 M111	Chapel Brampton, Northamptonshire	4/89	AR
4003	*Lode Star*	4-6-0	H6 M97	National Railway Museum, York (14)	3/92	SE
4073	*Caerphilly Castle*	4-6-0	H13 M94	Science Museum, London	6/61	SE
4079	*Pendennis Castle*	4-6-0	H13 M100	Hammersley Iron Mine, Australia (15)	7/77	RO
4110		2-6-2T	J28 M112	Southall Railway Centre	5/79	UR
4115		2-6-2T	J28 M112	Wales Railway Centre, Cardiff	2/88	AR
4121		2-6-2T	J28 M112	Dean Forest Railway	2/81	UR**
4141		2-6-2T	J28 M112	Dean Forest Railway (16)	See Note (16)	UR**
4144		2-6-2T	J28 M112	Didcot Railway Centre	4/74	UR
4150		2-6-2T	J28 M112	Severn ValleyRailway (17)	1/78	UR

No.	Present Name	Type	Page Ref.	Current Location	Date of Arrival	Status
4160		2-6-2T	J28 M112	West Somerset Railway (18)	4/90	UR
4247		2-8-0T	J38 M113	Cholsey & Wallingford Railway	4/85	UR*
4248		2-8-0T	J38 M113	Under restoration at Swindon	See Note (19)	UR
4253		2-8-0T	J38 M113	Pontypool & Blaenavon Railway	8/87	UR
4270		2-8-0T	J38 M113	Swansea Vale Railway, Llansamlet	7/85	AR
4277		2-8-0T	J38 M114	Gloucestershire & Warwickshire Railway, Toddington	6/86	UR**
4555		2-6-2T	J46 M115	Dart Valley Railway	10/65	RO
4561		2-6-2T	J46 M115	West Somerset Railway	9/75	R10/89
4566		2-6-2T	J46 M115	Severn Valley Railway	8/70	R7/75
4588		2-6-2T	J46 M115	Dart Valley Railway	11/70	R8/71
4612		0-6-0PT	E77 M84	Under restoration at Swindon	See Note (20)	UR
4920	*Dumbleton Hall*	4-6-0	H29 M104	Dart Valley Railway	6/76	R6/92
4930	*Hagley Hall*	4-6-0	H29 M104	Severn Valley Railway	1/73	R9/79
4936	*Kinlet Hall*	4-6-0	H29 M104	Gloucestershire & Warwickshire Railway, Toddington (21)	3/85	AR
4942	*Maindy Hall*	4-6-0	H29 M104	Didcot Railway Centre	4/74	AR
4953	*Pitchford Hall*	4-6-0	H29 M104	Dean Forest Railway	2/84	UR**
4979	*Wootton Hall*	4-6-0	H29 M104	Fleetwood (Lancs) Locomotive Centre	10/86	AR
4983	*Albert Hall*	4-6-0	H29 M104	Birmingham Railway Museum	10/70	UR
5029	*Nunney Castle*	4-6-0	H13 M100	Didcot Railway Centre	5/76	R3/91
5043	*(Earl of Mount Edgcumbe)*	4-6-0	H13 M99 M101	Birmingham Railway Museum	9/73	FS

No.	*Present Name*	*Type*	*Page Ref.*	*Current Location*	*Date of Arrival*	*Status*
5051	*Drysllwyn Castle* *Earl Bathurst***	4-6-0	H13 M101	Didcot Railway Centre	2/70	R1/80
5080	*Defiant*	4-6-0	H13 M101	Birmingham Railway Museum (note 22)	5/75	R8/87
5164		2-6-2T	J28 M112	Severn Valley Railway	1/73	R12/79
5193		2-6-2T	J28 M112	Southport Steam Centre	8/79	UR
5199		2-6-2T	J28 M112	See note (23)	See Note (23)	UR**
5224		2-8-0T	J38 M114	Great Central Railway, Loughborough	11/78	R11/84
5227		2-8-0T	J38 M114	Wales Railway Centre, Cardiff	2/88	AR
5239	*Goliath**	2-8-0T	J38 M114	Dart Valley Railway	6/73	R7/78
5322		2-6-0	J12 M110	Didcot Railway Centre See note (24)	9/73	5/73 (Note 24)
5521		2-6-2T	J46 M115	Dean Forest Railway (25)	12/80	UR**
5526		2-6-2T	J46 M115	Dart Valley Railway See note (26)	11/92	UR
5532		2-6-2T	J46 M115	Llangollen Railway See Note (27)	1/89	AR
5538		2-6-2T	J46 J50	Barry Island Promenade See Note (28)	4/92	SE
5539		2-6-2T	J46 M115	Wales Railway Centre, Cardiff	2/88	AR
5541		2-6-2T	J46 M115	Dean Forest Railway	10/72	R11/75
5542		2-6-2T	J46 M115	West Somerset Railway	9/75	UR
5552		2-6-2T	J46 J50	Bodmin & Wenford Railway	6/86	UR
5553		2-6-2T	J46 J50	An industrial site in the Midlands for restoration. See Note (29)	5/91	UR
5572		2-6-2T	J46 M115	Didcot Railway Centre (30)	7/77	R4/85
5619		0-6-2T	E75 M78	Telford Horsehay Steam Trust, Horsehay, Salop (31)	5/73	R4/81
5637		0-6-2T	E75 M78	Swindon & Cricklade Railway (32)	4/82	UR

No.	*Present Name*	*Type*	*Page Ref.*	*Current Location*	*Date of Arrival*	*Status*
5643		0-6-2T	E75 M78	Lakeside & Haverthwaite Railway (33)	5/89	UR
5668		0-6-2T	E75 M78	Pontypool & Blaenavon Railway	8/87	AR
5764		0-6-0PT	E77 M80	Severn Valley Railway	6/71	RO
5775		0-6-0PT	E77 M80	Keighley & Worth Valley Railway (34)	1/70	RO
5786		0-6-0PT	E77 M80	Bulmers' Railway Centre, Hereford (35)	7/70	RO
5900	*Hinderton Hall*	4-6-0	H29 M104	Didcot Railway Centre	6/71	R4/76
5952	*Cogan Hall*	4-6-0	H29 M105	Llangollen Railway (36)	6/89	AR
5967	*Bickmarsh Hall*	4-6-0	H29 M105	Pontypool & Blaenavon Railway	8/87	AR
5972	*Olton Hall*	4-6-0	H29 M105	Bombardier Prorail, Horbury Junction, Wakefield	5/81	UR**
6000	*King George V*	4-6-0	H20 M103	Stored at G.W.R. Museum Swindon (37)	3/92	RO
6023	*King Edward II*	4-6-0	H20 M103	Didcot Railway Centre (38)	See Note (38)	UR
6024	*King Edward I*	4-6-0	H20 M103	Didcot Railway Centre (39)	5/90	R4/89
6106		2-6-2T	J28 M113	Didcot Railway Centre (40)	11/67	RO
6412	*(Flockton Flyer)* *	0-6-0PT	E84 M88	West Somerset Railway (41)	3/76	RO
6430		0-6-0PT	E84 M88	Llangollen Railway (42)	See Note (42)	UR**
6435		0-6-0PT	E84 M88	Dart Valley Railway	10/65	RO
6619		0-6-2T	E75 M78	North Yorkshire Moors Railway	10/74	R12/84
6634		0-6-2T	E75 M79	East Somerset Railway	7/81	UR
6686		0-6-2T	E75 M79	Wales Railway Centre, Cardiff	2/88	AR
6695		0-6-2T	E75 M79	Swanage Railway	5/79	UR
6697		0-6-2T	E75 M79	Didcot Railway Centre (43)	8/70	RO

No.	*Present Name*	*Type*	*Page Ref.*	*Current Location*	*Date of Arrival*	*Status*
6960	*Raveningham Hall*	4-6-0	H29 M105	Severn Valley Railway See Note (44)	5/77	See Note (44)
6984	*Owsden Hall*	4-6-0	H29 M105	Herring Bros, Bicester,	10/86	UR**
6989	*Wightwick Hall*	4-6-0	H29 M105	Buckinghamshire Railway Centre, Aylesbury	1/78	UR
6990	*Witherslack Hall*	4-6-0	H29 M105	Great Central Railway, Loughborough	11/75	R8/86
6998	*Burton Agnes Hall*	4-6-0	H29 M103, M105 and Note (45)	Didcot Railway Centre	12/67	RO
7027	*Thornbury Castle*	4-6-0	H13 M101	Dart Valley Railway (46)	4/89	AR
7029	*Clun Castle*	4-6-0	H13 M100 M101	Great Central Railway, Loughborough (47)	8/91	RO
7200		2-8-2T	J42 M114	Buckinghamshire Railway Centre, Aylesbury	9/81	AR
7202		2-8-2T	J42 M114	Didcot Railway Centre	4/74	UR
7229		2-8-2T	J42 M114	East Lancashire Railway, Bury (48)	4/90	UR
7325		2-6-0	J12 M110	Severn Valley Railway (Ran as 9303, 7/92-3/93)	8/75	R7/92
7714		0-6-0PT	E77 M85	Severn Valley Railway	3/73	R10/92
7715		0-6-0PT	E77 M80	Buckinghamshire Railway Centre, Aylesbury (49)	1/70	RO
7752		0-6-0PT	E77 M80	Birmingham Railway Museum	6/71	RO
7754		0-6-0PT	E77 M85	Llangollen Railway	9/80	UR
7760		0-6-0PT	E77 M80	Great Central Railway, Loughborough (50)	4/90	R3/88
7802	*Bradley Manor*	4-6-0	H36 M107	Severn Valley Railway	11/79	R4/93
7808	*Cookham Manor*	4-6-0	H36 M107	Didcot Railway Centre (51)	8/70	RO
7812	*Erlestoke Manor*	4-6-0	H36 M107	Severn Valley Railway (52)	4/76	R6/79
7819	*Hinton Manor*	4-6-0	H36 M107	Severn Valley Railway	1/73	R9/77
7820	*Dinmore Manor*	4-6-0	H36 M107	West Somerset Railway (53)	3/85	UR**

No.	Present Name	Type	Page Ref.	Current Location	Date of Arrival	Status
7821	*Ditcheat Manor*	4-6-0	H36 M107	Llangollen Railway (54)	6/89	UR**
7822	*Foxcote Manor*	4-6-0	H36 M107	Llangollen Railway (55)	11/85	R4/88
7827	*Lydham Manor*	4-6-0	H36 M107	Dart Valley Railway (56)	3/73	R3/73
7828	*Odney Manor*	4-6-0	H36 M107	East Lancashire Railway, Bury (57)	3/91	R5/88
7903	*Foremarke Hall*	4-6-0	H29 M105	Swindon & Cricklade Railway	6/81	UR
7927	*Willington Hall*	4-6-0	H29 M106	Wales Railway Centre, Cardiff	2/88	AR
L.M.S. 8431		2-8-0	M43 M130	Keighley & Worth Valley Railway	5/72	R12/75
9303		2-6-0	J12	Severn Valley Railway (became 7325, 3/93)	8/75	R7/92
9400		0-6-0PT	E81 M86	Great Western Railway Museum, Swindon	4/62	SE
9466		0-6-0PT	E81 M87	Buckinghamshire Railway Centre, Aylesbury (58)	9/75	R3/85
9600		0-6-0PT	E77 M85	Birmingham Railway Museum	1/74	UR
9629		0-6-0PT	E77 M85	Forecourt of Marriott Hotel, Cardiff	1986	SE
9642		0-6-0PT	E77 M85 M130	Swansea Vale Railway, Upper Bank, Swansea (59)	4/89	R10/69
9681		0-6-0PT	E77 M85	Dean Forest Railway (60)	10/75	R9/84
9682		0-6-0PT	E77 M85	Southall Railway Centre	11/82	AR
—	*North Star*	2-2-2	B11	Great Western Railway Museum, Swindon	6/62	SE
W.T.5	*Shannon*	0-4-0WT	M130	Didcot Railway Centre (61)	1/69	R3/70
B.P.& G.V.2	*Pontyberem*	0-6-0ST	K214 M130	Didcot Railway Centre (62)	6/77	AR
P.&M.2		0-6-0ST	K261, M130 and N39. Arrived on Middleton Railway, 4/62, stored until 11/73, sold for scrap and cut up 11/73.			

GREAT WESTERN DIESEL RAILCARS

No.	*Current Location*	*Page Ref.*	*Date of Arrival*	*Status*
4	Great Western Railway Museum, Swindon	L19(2)	7/84 (63)	R5/74
20	Kent & East Sussex Railway	L20(2)	4/66	RO
22	Didcot Railway Centre	L20(2)	7/78 (64)	R by 8/68

GREAT WESTERN AUTO TRAILERS

No.	*Current Location*	*Page Ref.*	*Date of Arrival*	*Status*
38	Telford Horsehay Steam Trust, Horsehay, Salop	L17(2)	9/77 (65)	R7/87
92	Didcot Railway Centre	L17(2)	7/77 (66)	AR
160	Severn Valley Railway	L17(2)	9/69	Cut up, 1970
163	Cambrian Railways of Oswestry, Oswald Road Depot, Oswestry	L17(2)	1/76	RO
167	Dean Forest Railway	L17(2)	5/76	RO
169	Gloucestershire & Warwickshire Railway, Toddington	L17(2)	8/81	AR
174	Cholsey & Wallingford Railway	L17(2)	11/83	UR
178	Dean Forest Railway	L17(2)	12/79 (67)	RO
190	Didcot Railway Centre	L17(2)	8/70	R12/92
212	Didcot Railway Centre	L10	3/70	UR
225	Dart Valley Railway	L18(2)	10/65	RO
228	Dart Valley Railway	L18(2)	10/65	RO
231	Didcot Railway Centre	L18(2)	12/67 (68)	RO
232	Dart Valley Railway	L18(2)	/66	RO
(233)	British Railways Technical Centre, Derby, as "Test Car No. 1", Departmental No. ADW150375	L18(2)	—	RO
238	Dart Valley Railway	M125	10/65	RO
240	Dart Valley Railway	M125	10/65	RO

Notes:

In the "Status" column, "AR" is "awaiting restoration"; "FS" is "acquired for spares"; "RO" is "acquired in running order" but note that, particularly with the earlier arrivals, some of the locomotives are currently laid aside awaiting major overhaul; "SE" is "static exhibit" and "UR" is "actively under restoration". Those marked "UR**" are being restored partly or wholly at sites other than those given in the "Current Location" column.

In the "Present Name" column, the asterisk * shows names not carried as G.W. locomotives; of these, Nos. 7, 8 & 9 were named by British Railways in

1956, No. 823 had been *Countess* in G.W. days and the name *Earl of Berkeley* had been allotted to No. 3217 but the nameplates were never affixed. Two sets of nameplates are available for No. 5051, shown **, which are carried at different times, see Part 8 (2) pages H18/9. The four names given to Dart Valley locomotives were well chosen but were not acceptable to some members of the preservation movement as they had no traditional background. The plates were put on Nos. 1420/50 in the summer of 1976, on No. 1638 between 8/83 and 4/85 (they were removed in 1/90) and on No. 5239 in the summer of 1979.

(1) 426 Is to be restored as Taff Vale No. 85.
(2) 450 Has been restored as Taff Vale No. 28, in T.V. livery, see Fig. P36.
(3) 1338 Delivered to an open-air museum at Bleadon & Uphill Station 3/64 as static exhibit and moved to Didcot on closure of the Museum in 1987.
(4) 1363 Arrived at Totnes 8/64, transferred to Bodmin 5/69 where restored 5/70.
(5) 1450 Was bought from British Rail in running order and arrived on the Dart Valley Railway in 5/66. It was resold, complete with *Ashburton* nameplates, in 1992 but will remain at Buckfastleigh D.V.R. for the 1993 season.
(6) 1466 Was stored at Totnes from 1964 to 1967.
(7) 1638 Was bought from British Rail in running order and arrived on the Dart Valley Railway in 11/67. Its D.V.R. nameplates had been removed before resale.
(8) 2180 Was previously a static exhibit at Newton Abbot Station (page B40, Part 2).
(9) 2818 Was stored on behalf of Bristol Museum from 1965 to 1975.
(10) 2873 Bought as a source of spares has already been partly "cannibalised" .
(11) 3205 Arrived at Totnes in steam, 2/9/65, left the Dart Valley Railway in 2/67 for the Severn Valley Railway and was again moved, to the West Somerset Railway, in 3/87.
(12) 3217 Was bought as No. 9017 and was renumbered somewhere between 8/63 and 6/65.
(13) 3802 Arrived at the Plym Valley Railway 9/84 and left for Bodmin 2/90.
(14) 4003 Entered G.W.R. Museum, Swindon, 6/62, transferred on loan to York, 3/92.
(15) 4079 Bought from B.R. 5/64 and kept at locations including Southall, Didcot, Market Overton and Carnforth. Made last run Birmingham-Didcot 29/5/77.
(16) 4141 Delivered to Severn Valley Railway 1/73, sold to Dean Forest Railway and moved from S.V.R. to Swindon for restoration 4/89.
(17) 4150 Arrived at Parkend from Barry 5/74, moved to Severn Valley Railway 1/78.
(18) 4160 Left Barry 8/74 for Gloucester, where stored until 5/75, moved to Tyseley where stored 5/75-5/81, moved to Plym Valley Railway where stored 5/81-4/90, thence to West Somerset Railway where restoration started at last!
(19) 4248 Moved from Barry to Brightlingsea, Essex, 5/86, where some work was done; moved to Swindon by new owner for restoration with a view to hiring-out.
(20) 4612 Arrived on Keighley & Worth Valley Railway from Barry, 1/81, to be used as spares for No. 5775. Acquired by owner of No. 4248 (note 19) and also at Swindon for the same purpose. It will incorporate 5775's old boiler.
(21) 4936 Arrived on Peak Railway, Matlock, from Barry 5/81, moved to G.&W.R. 3/85.
(22) 5080 Left Barry by rail 9/8/74 and was stored at Gloucester until moved to Tyseley 24/5/75.
(23) 5199 Arrived at Toddington, G.&W.R., 7/85. Records conflict but most credible version is boiler only moved to Llangollen Railway 11/88 and most other components to Long Marston 12/90. To be offered on hire to Llangollen Railway on completion.
(24) 5322 Arrived at Caerphilly from Barry 3/69. First steamed 12/70 but first reported "in working order" 28/5/73. Moved to Didcot 9/73.
(25) 5521 Arrived on West Somerset Railway from Barry 9/75, left for Dean Forest Railway 12/80.
(26) 5526 Arrived at Toddington, G.&W.R., 7/85. Moved to Swindon Workshops for restoration 6/88 and to Dart Valley Railway 11/92 for completion.
(27) 5532 Arrived on Dean Forest Railway from Barry, 4/81, bought as a source of spares. The frames and cylinders moved to Llangollen Railway in 1/89 and were fitted with 5538's boiler and wheels. It will run eventually as 5532.
(28) 5538 Arrived on Llangollen Railway from Barry, 1/87. Boiler and wheels fitted to 5532's frames, as above, frames moved to Dean Forest Railway where fitted with 5541's tanks and wheels and sent to Steamtown 1/90 for preparation as

a static exhibit for display on Barry Island promenade, where it arrived in April 1992. No. 5541 in turn is to have the wheels recovered from No. 5532 and newly-made tanks.

(29) 5553 Originally intended for static display in Barry, was found in better state than 5538 and went to Dean Forest Railway 1/90. Moved to an unidentified Midlands industrial site for restoration, 5/91.

(30) 5572 Left Barry by rail for Taunton 8/8/71, moved thence to Didcot 7/77. In use 4/85 but finally painted and fitted with auto gear and steam heating 3/86.

(31) 5619 Is on long loan to the Gloucestershire & Warwickshire Railway, Toddington.

(32) 5637 Moved from Barry to Gloucester and then to Tyseley in company with No. 4160, see above. Moved from Tyseley to Swindon & Cricklade Railway 4/82.

(33) 5643 Moved from Barry to Cwmbran for abortive Eastern Valley Railway project 9/71, moved to Steamtown Railway Centre, Carnforth, by the end of 1972 (where some work was done), moved to Lakeside & Haverthwaite Railway 5/89.

(34) 5775 At first ran as L89 in London Transport livery, repainted in special livery for "Railway Children" film and more recently repainted as G.W. No. 5775.

(35) 5786 Arrived on Severn Valley Railway from London Transport 10/69 as L92, fully restored to G.W. livery as No. 5786 by 5/70, moved to Hereford 7/70.

(36) 5952 Arrived on Gloucestershire & Warwickshire Railway from Barry 3/81, moved to Llangollen Railway 6/89.

(37) 6000 Kept in Swindon Stock Shed until 8/68, overhauled by Adams, Newport, and first steamed at Bulmers' Railway Centre, Hereford, 13/11/68 and widely used on special trains. Returned to Swindon 4/90 and moved to G.W.R. Museum 3/92 pending decision on future use or static display.

(38) 6023 Delivered from Barry to Harveys of Bristol at Temple Meads Station 12/84. Some work done but loaded in sections to Didcot February & March 1990.

(39) 6024 Arrived at Buckinghamshire Railway Centre from Barry 3/73 where restored to full working order 4/89. Moved to Tyseley 10/89 and to Didcot 5/90.

(40) 6106 Bought from B.R. on withdrawal, stored at Southall 1/66-5/67 then stored at Taplow until moved in steam to Didcot, 4/11/67.

(41) 6412 Stored at Exmouth Junction from withdrawal until delivery in steam to Dart Valley Railway 5/6/66. Worked first public passenger train 5/4/69, moved in steam to W.S.R. 24/3/76. Name painted on for a children's T.V. programme in the late 1970's but does not appear on a photograph taken in February 1979.

(42) 6430 Bought from J. Cashmore Ltd. late 1964, stored at Exmouth Junction until 5/6/66 when towed to Dart Valley Railway by 6412. Intended as a source of spares for Nos. 6412/35 but sold about 1990. Exact dates not available but by early 1991 the boiler had arrived at Llangollen and the frames and wheels at Long Marston for restoration.

(43) 6697 Data incomplete but in Oswestry Works 9/66, arrived Tyseley in running order 10/67, reported at Ashchurch 8/68, in steam to Tyseley Open Day (28/9/69), left Ashchurch for Didcot in company with No. 7808, 8/70.

(44) 6960 Moved from Barry to Steamtown Railway Centre, Carnforth 10/72. Although it ran in steam at Shildon (page N3) on 31/8/75 the official restoration date is 5/76. Moved to Severn Valley Railway 5/77.

(45) 6998 Delivered in steam to Totnes Quay 4/66, moved Totnes to Didcot 12/67.

(46) 7027 Moved from Barry to Tyseley 8/72, moved to Dart Valley Railway, Buckfastleigh 4/89 but intended to be returned to Tyseley unrestored.

(47) 7029 Arrived Birmingham Railway Museum 3/66, moved to G.C.R. Loughborough 8/91.

(48) 7229 Left Barry for Plym Valley Railway 10/84, moved to East Lancashire Railway 4/90.

(49) 7715 Has run as "7715" and as "L99", see extended reference in text.

(50) 7760 Arrived Birmingham Railway Museum 6/71. Did not have its official steam test until 3/88 (date used in "status" column) but had been in light steam on a number of earlier occasions. Moved to G.C.R., Loughborough, 4/90.

(51) 7808 Data incomplete but worked from Birmingham to Taplow 17/9/66, reported at Ashchurch 8/68, in steam to Tyseley Open Day (28/9/69), left Ashchurch for Didcot in company with No. 6697, 8/70.

(52) 7812 Moved from Barry to Ashchurch by rail, calling at Parkend (D.F.R.) en route, in 5/74 and moved on to Severn Valley Railway 4/76.

(53) 7820 Moved from Barry to Gwili Railway 9/79, no work done and moved to W.S.R. 3/85.
(54) 7821 Moved from Barry to G.&W.R. Toddington 6/81, moved to Llangollen 6/89.
(55) 7822 Moved from Barry to Oswestry 1/75 where much work was done before transfer to Llangollen 11/85.
(56) 7827 Moved from Barry to Newton Abbot 6/70 where restored in under three years and left for Paignton and Kingswear in steam 30/3/73.
(57) 7828 Moved from Barry to G.&W.R. Toddington 6/81 where restored but made its first service run on transfer to Gwili Railway 5/88. Moved to Llangollen 3/89 and moved again to East Lancashire Railway 3/91.
(58) 9466 Temporarily renumbered 9404 for a short time in 1985, see text.
(59) 9642 Bought from R.S. Hayes' scrap yard, Bridgend, and delivered to N.C.B. Shed, Maesteg, 7/68, where put in running order 10/69. To store at Baglan, near Port Talbot 1984 and moved to Swansea Vale Railway 4/89.
(60) 9681 Ran disguised as No. 3775 on the Dean Forest Railway from July to October 1992.
(61) *Shannon*. Was formerly on static display at Wantage Road Station, see text.
(62) *Pontyberem*. Was formerly at Taunton, where it arrived from South Wales in 7/70.
(63) 4 Acquired from B.R. by Swindon Corporation. Information incomplete but was repainted in G.W. livery at Swindon in March 1960, was in store at Swindon, April 1967, became part of the National Collection, was on loan to Didcot Railway Centre 10/68-2/79 (made serviceable at Didcot May 1974), thence to York Railway Museum and to G.W.R. Museum, Swindon, in July 1984.
(64) 22 Arrived on Severn Valley Railway 5/67, where restored by 8/68 and moved to Didcot 7/78.
(65) 38 Moved from Cardiff to Ashchurch 2/72, then to Butterley 12/75 and to Horsehay 9/77.
(66) 92 Arrived from B.R. at Wills' Sidings, Swindon, 6/70, moved to Taunton 4/72 and to Didcot 7/77.
(67) 178 Arrived on Severn Valley Railway from Wolverton 1/69, to D.F.R. 12/79.
(68) 231 Bought 12/64 and stored at Totnes until 12/67, thence to Didcot.

MAIN LINE STEAM WORKINGS

These can be divided chronologically into three categories: the exploits of *City of Truro*, maintained by B.R. on behalf of the National Railway Museum; charter trains and other special workings by locomotives stationed on preserved railways or at private locomotive depots; and scheduled workings by private railway companies over B.R. metals into B.R. stations.

(1) No. 3440 *City of Truro*

The history of the "City" Class 4-4-0's appears on pages G36-8, Part Seven. Withdrawn in March 1931 as No. 3717, *City of Truro* went to York Railway Museum on 20th March and remained as a static exhibit until sent to Swindon, where it arrived on 11th January 1957, for overhaul and "for subsequent use for hauling special trains organised by Railway Societies". The overhaul included a boiler change and restoration to 1903 livery with its old number, 3440. It made running-in trips between Swindon, Didcot and Bristol in preparation for its first railtour duty on 30th March when it headed a Paddington to Porthmadog special between Wolverhampton and Ruabon. From mid-April it had a regular working (except when chartered) on the 12.42 Didcot-Southampton Terminus and 16.55 return, breaking this routine to work a Pontypridd to Swindon special with return to Cardiff on 23rd April and then the RCTS "North Somerset" rail tour from Reading on 28th April when it achieved 79 m.p.h. at Lavington, on the West of England main line. Later the same day it turned up at Paddington. On 18th May it worked, with No. 4358, to Hereford on Ian Allan's "Daffodil Express" and on 19th it reached Kingswear, worked an SLS special from Wolverhampton to Swindon and back on 16th June and ran to Llanelli on 2nd July where it had an argument with the coal stage. On 18th August it worked the eleven-coach "Moonraker", another RCTS special, from Paddington to Swindon and back and on 12th August it worked as pilot on the 5.30 Paddington-Plymouth, whence it worked a special to Penzance on 15th August, which ended its special duties for 1957.

1958 was another active year. Special trains included one from Swindon to Cardiff on 8th April and a Greenford to Horsted Keynes Ramblers' Special on 11th May and from mid-summer it frequently headed an up business train from Reading in the morning, returning on the 18.20 Paddington. Later that year No. 3440 was used as pilot or was in sole charge of the 17.32 Swindon-Bristol via Badminton, a duty it continued to work on occasions into 1960. It frequently worked Engineers' Saloons. On 26th August 1959 it journeyed to Scotland, to appear at the Scottish Industries Exhibition, following which it participated in several special workings in connection with the Exhibition. These covered double-heading with G.N.S.R. 4-4-0 No. 49 *Gordon Highlander* from Montrose to Glasgow and back on 3rd September, and again with No. 49 from Aberdeen to Glasgow and back on 5th and solo from Aberdeen to Glasgow on 9th but failed to make the return journey due to a heated big-end. This was put right in time for two return trips between Aberdeen and Glasgow on 15th and 19th September, paired with N.B.R. 4-4-0 No. 256 *Glen Douglas*.

Highlights of 1960 included two double-headed trips between Kings Cross and Doncaster, with Midland Compound No. 1000 on 20th and with the last B12/3 4-6-0 No. 61572 on 26th April. On 14th May it worked an enthusiasts' special over the M.&S.W.J.R. from Gloucester to Southampton Docks, returning over the former D.N.&S. line to Didcot (which had closed to passenger traffic on 5th March). On 4th September it headed an SLS special from Birmingham to Swindon and was on view at Birmingham Moor Street with Caledonian "single" No. 123 on 12th September.

The end of this first active period came in 1961. It visited the West Country between 20th and 25th March under charter to the Independent Television Company in connection with the Westward TV studio in Plymouth but was taken out of running stock on 27th May 1961 and was restored as a static exhibit, the number reverting to 3717, in time for the opening of Swindon Museum in June 1962. It was an obvious exhibit for the GWR 150 celebrations and left the Museum on 11th July 1984 for the Severn Valley workshops at Bridgnorth, where a protracted overhaul took until 3rd September 1985 when it emerged as No. 3440 again, in the 1897 Queen Victoria Diamond Jubilee livery. In the event it had no part in the GWR 150 celebrations and spent some time working advertised services on the S.V.R.

In 1986 it was back on the main line,

piloting No. 6000 *King George V* from Shrewsbury to Hereford on 24th May and working from Tyseley to York and back on 12th July, via Derby and Sheffield. On 1st August it ran light from Tyseley to York to join a pool of steam locomotives used on the "Scarborough Spa Expresses" and its last duty in 1986 was to work the Kings Cross-Scarborough "Dickens Festival Express" onwards from York.

The only reported workings in 1987/8 were again on the "Scarborough Spa Express" but 1989 was more eventful. On 23rd April, *City of Truro* was booked to work a Derby to Didcot tour but springs and an axlebox gave trouble and it was taken off for attention at Tyseley, leaving on 11th May for Didcot in time for a steam day on 14th May. The only other preserved G.W. 4-4-0 No. 3217 was on loan and they were pictured in tandem (Fig. P13). It left Didcot on 11th June for the "Trains Through Time" Festival in Utrecht, Holland, a celebration of 150 years of Dutch Railways (Fig. P14) and arrived back in York on 20th July, to return to Swindon in time for the opening of the "National Railway Museum on Tour" exhibition on 10th April 1990 when, as befits a V.I.P., was in steam for the Press. The exhibition closed on 4th November after which it was stored until moving to Bulmers' Hereford where it arrived on 23rd March 1991 and which was its base for the summer but made at least four forays: to Cardiff Cathays in time for the Taff Vale 150th Anniversary celebrations on 22nd June; to the S.V.R. for filming during the first week in July; to Didcot, which it reached by 15th August, leaving there just before midnight in a cavalcade with Nos. 5029, 6024, 6998 and 71000 to attend the Open Day at Old Oak Common on 17th August; and on 20th August it arrived on the Dean Forest Railway where the highlight was hauling the train for the official opening of Lakeside Station on 8th September. (This station was first used on 18th August, see Page P8). The following day it moved to the Severn Valley Railway and after some trips on that railway, left again for Bulmer's Railway Centre. It wintered here, earning its keep on the now traditional "Santa Specials" and then left for the National Railway Museum, York, where it was in steam in the Museum yard to mark the reopening of the restored Great Hall, on 15/16th April 1992.

On 2nd May it ran light engine to Butterley and worked a special train from Derby to Paddington the following day. A return trip, to be the last before expiry of its seven-year boiler certificate, should have run on 10th May but was cancelled through lack of support. Its next move was by road to Truro, where it was on show outside the Cathedral in its namesake city as part of Truro's Victorian Fair from 15th to 25th May. It then made several journeys over preserved railways, being on the Bodmin Steam Railway 26th May - 8th June, Dart Valley Railway (Totnes-Buckfastleigh) 9th - 29th June and West Somerset Railway thence until the end of July. Towards the end of August it was returned to York and put on static exhibition again but surely will be back on the active list in due course.

(2) Other Main Line Steam Workings

When the last public steam-hauled passenger trains ceased in August 1968, B.R. announced that steam locomotives would not be allowed to run on the main line except where agreements (as applied to No. 4079 *Pendennis Castle*) already existed. The absolute ban was soon lifted, witness the movement of Nos. 6697 and 7808 from Ashchurch to Tyseley and back in September 1970.

In 1971 the official attitude had softened further and permission was given for a series of trial runs by No. 6000 *King George V*, which at the time was stationed at the rail-connected Bulmers' Cider Factory, Hereford. Bulmer's owned a rake of Pullman cars, including an exhibition coach, known as the "Cider Train" and the following journeys were made in the early part of October 1971: On 2nd October Hereford-Swindon-Didcot West Curve-Oxford-Tyseley; 4th October, Birmingham Moor St.-Banbury-Olympia (where Bulmer's products were on view in the Exhibition Coach for two days) returning on 7th, Olympia-Reading-Swindon. No. 6000 was exhibited at its birthplace the following day and returned thence to Hereford on 9th October.

The success of this programme led to a relaxation of the steam ban and the first trip under the new dispensation was made by No. 7029 *Clun Castle*, which ran from Tyseley to Didcot and back on 11th June 1972. Stringent requirements were laid down and all locomotives had to be passed by B.R. inspectors before being allowed on the main lines and each train had to have a B.R. man on the footplate and a B.R. guard. Organisers had to make arrangements for watering and for turning; tender first running obviously would not be allowed and routes had to have either a reversing triangle or a turntable at each end; the latter of course existed only in preserved depots. This and the density of other traffic limited the available routes.

At this time locomotives passed for main line running were Nos. 1466, 4079, 6000, 6106, 6697, 6998, 7029 and 7752 (Nos. 5322 and 7760 were given in error on page M134, Part 12). Of these, only the 4-6-0's were really "main line" locomotives, used on the long-distance workings. No. 1466 had this status to enable it to run for five miles along the relief line from Didcot to Cholsey, to work on the Cholsey & Wallingford preserved line, while No. 7752 is believed to have worked special trains over the three miles between Tyseley and Birmingham Moor St., but also made a much more significant journey to the Stockton & Darlington 150th anniversary celebrations at Shildon in 1975 (an event also reached in steam by Nos. 6960 and 7808). No. 6106 ventured further afield and travelled to the Old Oak Common Open Days held on 2nd and 3rd September 1972.

These locomotives had been acquired in running order and relatively little needed to be done to fit them for main line running. All subsequent approvals of Swindon-built steam locomotives except No. 3440 had been restored from scratch by the preservation bodies and it is an outstanding tribute to the skill and dedication of those concerned that 4-6-0's Nos. 4930, 5029/51/80, 5900, 6024, 6960, 7808/12/19 and 75069; 2-8-0's Nos. 2857 and 3822; 2-10-0's Nos. 92203/20; 0-6-0PT Nos. 7760 and 9466 and 2-6-2T Nos. 4566 and 5572 have been added to the list. Boiler certificates for main line running last for only seven years and of the foregoing, only Nos. 4566, 5029/80, 6024, 6998, 7029, 7752/60 and 75069 still have valid certificates, while Nos. 9303 and 46521, both of the Severn Valley Railway, were certificated in 1992. At the time of writing Nos. 1466, 6000 and 6697 are laid aside for major overhaul and No. 4079 has of course gone to Australia.

Reverting to 1972, there still were official misgivings about failures on the principal main lines for when No. 6000 and its train ran from Hereford to attend the Old Oak Common Open Days of 2nd & 3rd September, and from Hereford to Plymouth for Laira Shed's Open Day on 23rd September, No. 6000 was allowed only in "light steam" and was piloted by a "Hymek" diesel locomotive which obviously did most of the work. No. 6000 *was* allowed to work a round trip between Newport and Shrewsbury on its own on 16th September, because a failure would have had less drastic repercussions.

A pattern of operating was established, which has continued up to the present, whereby the bulk of the special workings have been sponsored by the principal preservation centres at Bridgnorth, Didcot and Tyseley, by Railway Societies, by railway book and magazine publishers and by travel agents (some set up for this purpose). Most of the trips ran but there were the inevitable cancellations due to lack of support or of non-availability of locomotives or carriages at the critical time. Locomotives have failed en route due to bearings running hot or to bad steaming caused by poor quality coal, tube failures and so on but with the resilience typical of the steam locomotive, journey's end often was reached late rather than the indignity of being rescued by a diesel locomotive. During the last few dry summers there has been a steam ban on some occasions to avoid the risk of lineside fires, with substitution of diesel haulage or cancellation.

The foregoing activities have been fully reported in the contemporary *Railway Observers* and elsewhere and will not be described in detail but it is appropriate to describe steam workings sponsored by the Western Region of B.R., as successors in title to the G.W.R. Except for the *City of Truro* workings, described earlier, these trains had to be worked by locomotives hired from the preservation bodies, and fell into two categories, celebrations and commercial ventures. By far the most spectacular was the 150th Anniversary Celebrations, which justify a detailed account.

To commemorate the giving of the Royal assent to the Great Western Railway Bill on 31st August 1835, under which the Great Western Railway was incorporated, Western Region prepared a comprehensive programme of events which included an Exhibition Train, to be stationed at various G.W. locations, a static exhibition of locomotives, rolling stock and machinery at Swindon Works, special naming ceremonies of modern motive power, record-breaking high speed runs and last but not least, the running of main line excursions hauled by steam locomotives. The events were planned to run from 3rd April to 9th November, appropriately starting with the replica 4-2-2 *Iron Duke* running in steam with replica carriages on broad-gauge track in front of the Albert Memorial in London's Kensington Gardens and ending with a G.W.R. 150 Float in the Lord Mayor's procession in London on 9th

November. The preserved Steam Railways and Museums which are closely associated with the G.W.R. also celebrated 150 years of Great Western progress in appropriate style with special trains and exhibitions.

Unhappily the programme was not without its troubles, getting off to a bad start by the announcement of closure of Swindon Works at the worst possible moment. The staff understandably had nothing to celebrate and the Swindon Exhibition was abandoned. The gloom was compounded by locomotive failures at the worst possible time, which first affected the Bristol to Plymouth "Great Western Limited" on 7th April. This was double-headed from Bristol by Nos. 6000 and 7819 but No. 6000 had to be detached at Taunton with a hot tender axlebox. No. 7819 was assisted at the rear of the 13-coach train by two Class 37 diesel locomotives as far as Tiverton Junction but No. 7819 also developed a hot tender axlebox and had to be removed at Exeter, the two diesel locomotives continuing to Plymouth. After repair, No. 7819 proceeded to Plymouth but No. 6000 needed more attention and had to be replaced on the return journey on 8th April by No. 4930 which had been despatched at very short notice from Bridgnorth, Severn Valley Railway. Indifferent running was experienced through poor quality coal but otherwise there were no problems. The "jinx" also affected the "Great Western Limited" of 7th July, which left Bristol behind Nos. 5051 and 4930 but a track circuit failure at Aller Junction caused a slow start for Dainton bank and, due also to poor coal, the train stalled on the bank. Class 50 diesel No. 50045 assisted to Totnes where Nos. 5051 and 4930 were detached. These two should have worked from Plymouth to Bristol on 14th July but took over from Newton Abbot.

The other trains in the programme ran as planned (except that there were no reports of the scheduled Bristol-Plymouth "Great Western Limited" on 1st September). In date order these were:

4th June: G.W.R. Exhibition train, Didcot to Tyseley, No. 5051, the only recorded occasion when this train was steam-hauled.

8th & 9th June: "Shakespeare Express" from Birmingham (Hall Green on 8th, Tyseley on 9th) to Stratford-upon-Avon, No. 7029.

27th July: Cardiff-Swindon-Oxford and back, worked by Swindon-built B.R. locomotives Nos. 75069 and 92220.

6th, 7th, 11th, 13th and 14th August: Two return trips each day between Swindon and Gloucester, worked variously by Nos. 4930, 5051 and 7819. Similar trips were run on 18th, 20th, 21st, 25th, 26th, 27th & 31st August and 1st September using Nos. 6000, 7029 and 75069.

No. 7029 visited the West Country in September, working a Plymouth-Par excursion on 6th, appearing with No. 5051 at Laira Shed's Open Day on 7th; on 8th September the two "Castles" worked the last "Great Western Limited" between Plymouth and Bristol. On 10th September, No. 7029 worked the "Samuel Whitbread" special from Newport to Gloucester and back, while 2-8-0 No. 2857 worked a "Railfreight Spectacular" of 26 preserved freight wagons from Newport Alexandra Dock Junction to Newport Station. The series was rounded off by three return trips between Swansea and Carmarthen, worked by No. 6960 on 21st September. No. 6960 also worked two trips the following day, with the third in the hands of No. 5051.

There are no reports of W.R.-sponsored steam trains in 1986 but new ground was broken in 1987 by the "Cardigan Bay Express" over part of the Cambrian system. Nos. 7819 and 75069, both types which had worked the route in steam days, arrived from the S.V.R. at Machynlleth on 21st May and the "Manor" worked the publicity special to Pwllheli the following day. There were no turning facilities so No. 7819 ran, chimney first, to Barmouth, on 24th-27th May, while No. 75069 left Machynlleth tender-first on 28th, 29th and 31st May. Further trains operated between 20th July and 2nd September but 75069 was unavailable and the "Swindon" atmosphere was reduced as S.V.R.'s 2-6-0 No. 46443, built at Crewe had to deputise for the 4-6-0. There was at least one working between Machynlleth and Aberystwyth utilising, and deputising for, a morning "Sprinter" shuttle path between these stations, a rare case of a scheduled passenger service being steam-hauled.

Upgrading of the track between Cardiff and West Wales to take heavy oil tank wagons made it possible to run "King" Class locomotives. As a result, No. 6000 (Fig. P1) joined No. 7029 in working the "Carmarthen Express" between Swansea and Carmarthen on four weekends in September 1987. These two should have worked the "Jubilee Express" between Swindon and Gloucester on 16th and 17th May, to celebrate the 25th anniversary of Swindon Museum, but No. 6000 managed to derail itself at Gloucester shed before the first trip and No. 7029 hauled all the trains. To mark the last day of operation from Birmingham Moor Street, a special hourly service of steam trains was operated thence to and from Dorridge on 26th September, using Nos. 7029 and 46443.

Apart from workings by *City of Truro*, see elsewhere, there do not seem to have been any W.R. steam services as such in 1988. Although the Welsh services had been successful, the market was considered too limited to repeat the ventures in 1988. A special train was worked from Kemble to Newport by No. 7029 on the evening of 7th May 1988, which seems to have been chartered by a party of agents and others who had attended the Badminton Horse Trials.

Great Malvern Station had been badly damaged in a fire in 1986 and as it is a Listed Building of high quality, much care was taken in restoration, culminating in the fixing of a plaque unveiled by (the first) Sir Robert Reid, on 20th May 1988, and who arrived in style on a train hauled by No. 7819. This was rather theatrical as the locomotive had been attached at Ledbury, only seven miles away and handed over to diesel traction again at Great Malvern, No. 7819 then running to Tyseley for wheel turning.

Reverting to 1986, the "Private Sector" worked a number of routine trips which are amply reported elsewhere but it is worth mentioning that No. 4930 worked special trains between Andover and Ludgershall, Southern Region, on 22nd and 23rd March, while No. 6998 was used on the "Blackmore Vale Express" specials between Salisbury and Yeovil on 5th, 12th, 18th and 19th October. 1987 was routine but an unusual working in 1988 was of 2-6-2T No. 4566 on shuttle services between Lever Brothers' Margarine Siding and Merseyrail's Port Sunlight station on 1st and 2nd May 1988.

2-8-0 No. 3822 made its only recorded main line movement when it took a train of four restored goods wagons and a brake van to the Electricity Board's Open Day at Didcot Power Station on 11th June 1988 and on 14th September H.R.H. Prince of Wales rode on the footplate of No. 7029 from Birmingham Snow Hill station to Tyseley, inaugurating the "London and Birmingham 150 Gala" held at Tyseley on the 17th & 18th of that month.

There were no reported W.R.-sponsored workings in 1989 and as from 1st January 1990, a partnership with the private sector was set up, comprising B.R.'s InterCity Sector, Flying Scotsman Services (FSS) and the Steam Locomotive Operators Association (SLOA) to handle future steam workings. 1990 also was marked by an announcement in Parliament that steam locomotives would be exempted from pollution controls in the Environmental Protection Bill.

The 1989 steam workings commenced with a "Royal Train" from Tyseley to Birmingham Snow Hill station on 14th January to commemorate the signing of the charter conferring "City" status on Birmingham on 14th January 1889. The train was hauled by No. 5080, with *City of Birmingham* nameplate on one side only. This had been carried by the "City" Class 4-4-0 and had been loaned by the Birmingham Museum of Science & Industry. 0-6-0PT No. 9466 was used to haul a series of special trains of B.R. stock between Chesham and Watford on 1st, 2nd, 8th & 9th July, in connection with the centenary of the opening of London Transport's Chesham Branch.

The 1990 steam workings followed the usual pattern but there was a welcome return to the Cambrian Coast line in 1991, albeit on two days only in June and two days in September. The locomotives were Nos. 7819 and 75069, as in 1987. On a minor key, it was announced that Nos. 7752/60 had been approved "for travel over B.R. lines to open days and similar events", presumably a renewal of 7752's "permit".

Steam specials between London and Stratford-upon-Avon had always started from Marylebone Station but in 1991 permission was given to use Paddington and the first train to do so ran on Saturday 9th November 1991, when the train was hauled by No. 5029, over 26 years since the previous occasion (referred to elsewhere) when No. 7029 worked the last scheduled steam train, on 12th June 1965. (There had been a steam locomotive working of sorts when No. 6000 headed a railtour on 1st March 1979 to celebrate Paddington's 125th anniversary but managed to run hot and had to be replaced by a diesel locomotive at Didcot).

1992 started with three round trips between Didcot and Oxford, hauled by Didcot's No. 6998, on 29th February, sponsored by Network SouthEast. On 14th March, No. 6024 worked from Paddington to Stratford-upon-Avon, announced as the last steam working from Paddington, yet *City of Truro* (see elsewhere) arrived there from Derby on 3rd May. To conclude this account, the BR/FSS/SLOA partnership is already planning for main line steam operation into the next century! With this in mind, B.R. certification procedures have commenced on Nos. 4121/41, 4953/83, 5521/6, 6695, 7715, 7802 and 92212/14, all of which are in various stages of restoration with No. 7802 virtually ready to run at the time this list (as at November 1992) was published in STEAM RAILWAY; the compiler is indebted to the Editor for permission to publish the list.

(3) Private Railway Companies' Trains and British Rail Stations

Private railways which start from or near to a main line station are more attractive to potential visitors if connections can be made with B.R. trains. In the case of the Bodmin & Wenford, Cholsey & Wallingford and Swindon & Cricklade Railways, cross-platform interchange is or soon will be possible and will apply to the Dean Forest and Llangollen Railways when their lines extend to Lydney and Ruabon respectively. The Severn Valley has its southern terminus very near Kidderminster B.R. and the Vale of Rheidol trains actually start from Aberystwyth station. The Dart Valley and West Somerset trains need to run over B.R. tracks to reach Totnes and Taunton stations respectively.

The Dart Valley made history when its first public passenger train ran into Totnes B.R. station on 5th April 1985, hauled by No. 1638 *Dartington.* At different times, Nos. 4555 and 4588 also were passed to run into Totnes, where the locomotive had to run round its train (Fig. P10). The service lasted for four summers but continually increasing costs associated with the use of B.R. facilities made it uneconomic and D.V.R. trains were diverted to the D.V.'s own station, Littlehempston, on the other side of the River Dart, from the 1989 season onwards.

SERVICES WORKED BY G.W.R. STEAM RAILMOTORS AND AUTO TRAINS

Revising pages N16-8 of Part 13, although the railmotor allocations and transfers between March 1906 and April 1908 still have not been found, Pendon Museum has kindly provided a list of Steam Railmotor and Auto Train workings in the summer of 1911 and which is reproduced as Table 2A. Coupled with research into services described as "Rail Motor Cars One Class Only" in Bradshaw's Railway Guide for April 1910, it becomes possible to confirm, clarify or correct much of the information given and assumptions made in Part 13 and the 1911 list gives a much broader picture than that in Table 2 on pages N16-8. Since Part 13 was published there also has been a spate of well-researched books on Great Western branch lines and some of the items on pages N16-8 should be altered, as below.

On a general note, several branches in the London Division, including Bourne End to Marlow, Cholsey & Moulsford to Wallingford and Radley to Abingdon would seem to have been suitable for steam railmotor working but the Wallingford branch was shown as "1st and 3rd Class" and the others may have had freight trips, worked in gaps in the passenger services which made for the use of small locomotives. Steam railmotors did haul tail traffic apart from conventional passenger trailers but were not ideal for shunting. Princes Risborough to Watlington, described as "One Class Only" was a more obvious case for operation by locomotive and carriages. This no doubt also accounts for the end of steam railmotor working on the Ashburton to Totnes branch and the absence of such weekday workings between Kingsbridge and Brent, although the latter did have steam railmotor services on Sundays, provided by Plymouth (see Table 2A).

Photographic evidence has come to light about multiple-unit working of steam railmotors, between Craven Arms and Wellington (Fig. P16), from Newnham to the Forest of Dean Branch (Fig. P17) and between Plymouth and Saltash (Fig. P15). A two-car train was no more difficult to control than railmotor and trailer but the method of controlling the regulator from the "trailing" end involved under-floor rodding and universal joints, which proved too cumbersome in practice with three-car and four-car formations. It appears that in these cases the rodding was disconnected and the regulators at each end operated separately; either a second driver had to be employed or a qualified fireman might have been allowed to control the trailing railmotor, instructed by bell-code signals from the leading driver.

The dates in the column headed *Period* on pages N16-8 were the start of the first known allocation and of the end of the last known allocation. The available information has now been plotted in bar-chart form which reveals that there were significant gaps in allocations which might have been expected to be continuous. The shorter gaps may reflect shortage of steam railmotors, which did seem to spend a lot of time in Swindon Works, but the longer gaps suggest either that the services were worked by other sheds; by auto or conventional trains; or that railmotor services as such were discontinued. There also is partial evidence of gaps where a railmotor seems to have been stored for periods of two to eight months before official transfer to the next depot but this does not affect the allocation periods at the depots involved. Where dates in Table 2 are affected they are included in the relevant notes below.

Working details for the Bristol Division for the summer of 1907 have been found, covering Chippenham, Frome, Taunton, Trowbridge and Weymouth, all of which had known allocations in 1906 and 1908. There was one working by Bath shed, and a curious Mondays-only working by Swindon (which also applied in 1911, see Table 2A) which looks like a regular running-in turn for railmotors ex Works.

Photographs have come to light of the boiler being craned into railmotor No. 73 at the makers' works (Fig. P18) and of an engine unit under overhaul in Swindon "A" Shop (Fig. P19). The railmotor bodies seem to have been maintained in the Carriage Shops.

Experimental services started in the 1920's to counter road motor competition were mostly unsuccessful, mainly because the buses both ran more frequently and, in relation to the average journey length, provided more of a door-to-door service. One such service was put on between Wolverhampton, Wombourn and Stourbridge Junction as from 11th May 1925 following the opening of a new through line between these places and Wolverhampton received an allocation of

REPLICATION AND PRESERVATION

Photograph] [R. J. G. Antliff

The replica *Iron Duke* on mixed gauge track laid entirely with genuine broad gauge rails and also incorporating a few broad gauge wooden baulks, at Didcot Railway Centre, on 2nd July 1986. P12

Photograph] [Peter Treloar

The two preserved G.W. 4-4-0's at Didcot Railway Centre on 12th May 1989. P13

Photograph] [Richard Gibbon

***City of Truro* at the "Trains Through Time" Festival at Utrecht, Netherlands in the summer of 1989.** P14

MULTIPLE UNIT WORKING OF STEAM RAILMOTORS

Photograph] [M. G. D. Farr's Collection

Saltash to Plymouth service worked by No. 7 and another steam railmotor working in multiple. P15

Photograph] [G. M. Perkins from HMRS Collection

Wellington service at Craven Arms in 1906, worked by No. 74 plus trailer and another steam railmotor. Note also Bishop's Castle 0-4-2T No. 1, ex G.W. No. 567 (Part 6, page F19 and Fig. F30). P16

Photograph] [Lens of Sutton

Two trailers sandwiched between two steam railmotors at Upper Soudley Halt on the Forest of Dean branch. P17

THE BEGINNING, THE END AND HARD WORK

This official photograph by the Gloucester Railway Carriage & Wagon Company shows the boiler being lowered into newly-built Railmotor No. 73 in April 1906.

Reproduced by permission of Messrs O.R. & J.R. Gibbon

P18

Photograph] [Ken Davies

Railmotor Engine Unit No. 0890 in Swindon "A" Shop, June 1935, having been condemned with Car No. 37. P19

Photograph] [K. Beddoes' Collection

Steam railmotor ascending the 1 in 40 gradient at Farley Dingle, between Buildwas and Much Wenlock. This "ruling gradient" would have precluded trailer haulage between Wellington and Craven Arms. P20

STEAM RAILMOTORS AT WORK FAR AND WIDE

Photograph] [Lens of Sutton

Railmotor No. 9 at Cynwyd (between Bala and Ruabon) before 1913.
P21

Photograph] [K Davies' Collection

Plymouth-Saltash railmotor crossing the Royal Albert Bridge, Saltash (reproduced from a card postmarked 1908). P22

Photograph] [National Railway Museum

Railmotor No. 85 and clerestory-roofed control trailer on Reading-Didcot service near Pangbourne. P23

two cars for this service, while Stourbridge shed had its allocation of four increased to five. The timetable was later recast to allow it to be worked by a single car based at Stourbridge but the service still was unviable; in 1931 its costs were said to be over seven times the receipts and it was discontinued as from 31st October 1932.

The foregoing, and other information which has come to light, involves the following amendments to the information given in Part 13, as follows:

Page N16, Column 1, paragraph 3: Delete Chester and Whitland. As shown in Table 2A these sheds had steam railmotor services in their own right.

Ditto, paragraph 4: A Brynamman to Pantyffynnon service was reported on page 435 of the *Railway Magazine* for 1905, which would account for the Brynamman allocation. The Llanidloes railmotor worked from there to both Moat Lane Junction and Builth Wells.

Ditto, paragraph 7, continued into Column 2: Delete entirely, see above.

Column 2, paragraph 5: Delete lines 4 & 5; the services started on 1st May 1906 and are reported worked by locomotive and carriages from some time in 1907. The gap in available official allocations covered the period and from the published timetable it appears that two cars were stationed at Wellington and two at Much Wenlock. The through workings between Wellington and Craven Arms were supplemented by several shorter trips and some cars ran on the main line to and from Shifnal, reversing there to and from the cross-country route. Traffic outgrew the capacity of the single cars (Fig. P20), 1 in 40 gradients precluded the use of trailers and "multiple units" comprising two power cars with a sandwiched trailer had to be introduced (Fig. P16), an uneconomic way of working leading to the change to locomotive and carriages.

Paragraph 5, lines 6-12: Table 2A confirms that these services were worked by Banbury auto trains with no involvement of Oxford railmotors.

Ditto, paragraph 7: The pioneer steam railmotors Nos. 1 and 2 were hired to the independent Lambourn Valley Railway from 15/7/04 to 1/7/05 (when the line was taken over by the G.W.R) because the traffic did not come up to expectations and did not justify locomotive haulage (see Part 3, page C91). Surprisingly, in view of the supposed shortage of passengers, Cars Nos. 19 and 21 went there later in 1904 and for a couple of months there was an allocation of four cars. Nos. 1 and 2 remained on the line in Great Western days, until January 1906 and October 1905 respectively.

Amend Table 2 on pages N16-8 as follows:

Ashburton: Delete "1". The allocation ended in 5/05. It is not clear whether a Newton Abbot car worked a service after 5/05 but it had disappeared by 1911, see Table 2A.

Bath: Delete "7/08". A single car was rostered in the summer of 1907. There were gaps in the allocation 5/16-4/17 and 1/24-8/25.

Bristol: The Clifton Down service was run at holiday times to relieve pressure on the Bristol-Avonmouth service. Between 7/13 and 5/15, Car No. 87 is shown as shared between Bristol, Yatton and Radstock.

Chalford: Readers have recollections of the Chalford-Stonehouse service being worked to and from Gloucester; there certainly was a Gloucester working on Saturday nights in 1925/6, which extended also at the other end to Kemble. Gap in allocation 12/18-7/21 and some short gaps also.

Cheltenham: Gap in allocation 12/11-4/13.

Chester: The 1911 workings appear in Table 2A, when Chester had only one car. In mid-1914 there were three cars and four in 1917, dropping back to two in 1920. The Hooton to West Kirby branch was an obvious subject for railmotor working but it is not known if they were used here.

Chippenham had two long gaps in allocation, 10/10-12/16 and 4/17-9/19.

Corwen: Although Table 2A shows workings to Wrexham and Oswestry from this shed, the Swindon allocation sheets show no car here until May 1912, apart from one in 1905/06. Perhaps the car was outstationed informally and which could account for the presence of No.9 at Cynwyd, en route for Bala (Fig. P21) as this was at Croes Newydd at various times. The Bala service is believed to have ended in July 1917.

Croes Newydd: In 1910 this shed, in common with Southall and Stourbridge, had an allocation of eleven cars and they shared one-third of the entire fleet. The list in Table 2A amplifies that in Table 2; the Oswestry service ran via Ruabon and Gobowen.

Didcot: The duties shown in Table 2A were designed to be worked by one car although a second was allocated for short periods and a spare was supplied by Oxford when needed.

Duffryn Yard: Cars also ran on the "main line" between Port Talbot and

Blaengarw. Allocations found for P.T.R. Car No. 1 show that it started work in 5/07 and ceased in 8/15, extending the "Period" at both ends.

Evesham's duties had ceased by the time Table 2A applied.

Exeter's allocation had one significant gap, 8/12-8/16.

Frome lost its allocation 7/22-2/24, balanced by an allocation to Westbury.

Garnant lost its allocation 8/17-10/21.

Gloucester: There were gaps in the allocation 9/14-12/16 and 9/19-7/20. Although Fig. P17 shows a four-car "multiple unit" working the Forest of Dean branch, the bulk of the work seems to have been done by auto trains. These started to Steam Mills Crossing Halt in 8/07, by 1910 had been extended to Drybrook Halt, and at a date unknown, served Cinderford by double reversal en route. There is no information on other Gloucester duties. Perhaps it was a "parent" shed for Chalford and Cheltenham.

Helston had five widely-spaced allocations totalling about 16 months in 8½ years. Table 2A shows a "Helston" car at a time when there was no official allocation and it seems as if the branch was worked in conjunction with services to Truro and Penzance.

Kidderminster: Amend "Services worked" to read "Kidderminster to Bewdley, Stourport and Hartlebury, returning main line to Kidderminster and vice versa." (Fig. P24). The Woofferton service shown in Table 2A was at one time worked by a Stourbridge car which ran empty between Stourbridge and Kidderminster.

Laira: The Plymouth suburban services developed so much traffic that they outgrew the capacity of the railmotors, even with multiple-unit working, and they were replaced by auto trains, often in four-car formation. The 1905 allocation of ten had dwindled to six by 1910 and Laira had none after January 1914 except for a couple of short-term allocations between 2/29 and 8/30. The Sunday service to Kingsbridge (Table 2A) ceased in 1912. (The original of Fig. P22 is postmarked 1908).

Llanelly's allocation applied only in 4/08-6/08, 5/20-5/21 (including the L.& M.M. branch trials mentioned on page N16) and 1/26-1/27.

Malvern Wells: A single car, No. 71, was allocated here between 8/17 and 1/20 but its duties have not been identified.

Merthyr: The railmotors to Newport would have used the G.W. route as far as possible, over the joint line to Quakers Yard (High Level), thence on the line to Pontypool Road as far as Maesycwmmer, then over the L.&N.W.R. line to Nine Mile Point, whence there were running powers to Risca and a G.W. line on to Newport. There were short trips between Merthyr and Quakers Yard.

Monmouth Troy: Car No. 22 was here from 7/11 to 10/11, on the duties shown in Table 2A and Car No. 80 was here from 5/14 to 8/14. Amend "Period" accordingly.

Neath: There were gaps in the allocation, 8/16-6/17 and 1/19-6/20.

Newport: The allocation varied from three to five cars but no details are available of the services worked.

Newton Abbot: Delete Ashburton to Totnes and Kingsbridge to Brent from "Services worked". There were gaps in the allocation, 1/09-11/10 and 5/15-4/16.

Newquay: The "period" extended from 2/05 to 5/11 (not 7/10), with workings as described in Table 2A.

Oswestry (and Machynlleth) received allocations in 8/22, at about the time that the Cambrian Railways became part of the G.W.R. Oswestry cars worked to Wrexham Central via Ellesmere but did not work an obvious Oswestry-Gobowen shuttle. Machynlleth seems to have been a "parent" shed for Penmaenpool (which see) but there is a photograph of No. 39, an Oswestry car, working between Dolgelley and Barmouth in August 1922.

Penmaenpool cars worked a basic service between Dolgellau (then spelt Dolgelley) and Barmouth. In the winter 1923 timetable the service was extended northwards to Llanbedr & Pensarn and was further extended on to Criccieth and southwards to Aberdyfi (then spelt Aberdovey) in the summer 1924 timetable. The allocation shows one car only but there seems to have been overlapping with Machynlleth.

Penzance had an allocation of three cars in 1904/5, two at the end of 1905 and one in early 1906, possibly continuous until 1908 and in most of 1909. There were two at the end of 1909 / early 1910 then one again until mid-1912 but with no duties at the time according to Table 2A. One in late 1912 increased to two and then three in 1914, further increased to three in mid 1915 and then back to two up until mid 1916 with a single car for part of 1921 and part of 1922. Although there are no references to turns worked by Penzance cars they must have been tied up with the services on the main line to Truro and with the Gwinear Road to Helston branch and probably other branches nearer Penzance: particularly as Truro lost its allocation in

June 1913 and did not regain any until 6/19, having one car only up to 4/21 when it went to Penzance.

Pontypool Road: There was a long gap in the allocation, between 2/17 and 6/23.

Radstock's allocation applied only between 12/10-8/11 and 6/14-4/15 but see "Bristol".

Reading: There was no allocation until 7/18 which replaces 5/14 in the "period" column. One of its turns is shown on Fig. P23 which also shows one of the clerestory trailers Nos. 14-8 & 35, described on page L17, Part 11, Second Edition.

Slough: There was a long gap in the allocation between 1/20 and 6/29.

Southall: This shed had the most complicated rosters of all, which needed an article in five instalments in the *Great Western Railway Magazine* of 1908 (and this was of course down to 1908 only) to describe their ramifications. As will be seen from Table 2A they worked on to other Companies' lines, serving Willesden Junction, Kensington, Clapham Junction and Victoria. Ten of the first 29 cars built were sent to Southall and some at first were outstationed at Westbourne Park but this was found to be too congested and all trips then started and ended at Southall, For the same reason, services terminated at Westbourne Park station on weekdays, only running to and from Paddington Station on Sundays; Fig. P25 shows car No. 89 leaving for Greenford at 5.05 p.m. and there was one trip as far as High Wycombe on Sundays but by 1911 (see Table 2A) cars did not run beyond Gerrards Cross. One turn, leaving Southall at 5.15 a.m. and arriving back at 10.18 p.m. covered a total of 218 miles in seventeen hours, an average of 13 miles an hour on stopping trains, including turnround and servicing time.

Stourbridge: The service to Stourbridge Town was worked without a guard (there may have been other cases over short branches with no intermediate stops). Table 2A shows services to Birmingham (Snow Hill) and to Hagley.

Stratford-upon-Avon: The 1908 allocation of two built up to a peak of four in 1911/2, dropped to three in 1913-15 and back to two in 1916/7. Table 2A gives a general picture of the services worked by this shed.

Swindon: Car No. 10 was here 10/05-3/06 (and possibly longer; no car was shown here in 1911 and the Mondays-only working of Table 2A could have been an ex-Works running-in turn). There were two cars 12/18-8/19. (Omission from Table 2, Part 13).

Tenby: Delete 6/08 to 9/22 in "Period". There were three cars in the summer of 1908, one in summer 1909 and then no allocation until 12/18 when again there were three cars, reduced to one in 1919 and ceasing entirely in 2/20. "Services worked" should be altered to read "Pembroke Dock Branch."

Trowbridge: Table 2A gives a better idea of the extensive services, which were maintained by an allocation of up to six railmotors.

Truro's allocation was not continuous, but there were five cars (not one, wrongly shown on page N18) in 1908, six for most of 1909, reducing to three in 1910/11 with an extra car for a short time in 1912 and ceasing in 6/13. Allocations resumed with two cars towards the end of 1919 with one car on until April 1921. They would seem to have been used on the main line to Penzance, with intermediate trips from both ends to Camborne and Redruth and extra trips over the three and a half miles between these places, which looks like a half-hearted attempt to combat tramway competition.

Tyseley received Car No. 85 in 5/08 and Nos. 86-8 in 6/08, perhaps for crew-training in advance of services on the North Warwickshire line which opened to passengers on 1/7/08. The allocation peaked at ten in 1910 but ended up at two in 1914.

Weymouth: It has been suggested that steam railmotors also worked the Maiden Newton to Bridport branch on Sundays only. There were gaps in the allocation 2/17-7/20 and 9/21-7/23.

Whitland: All the allocations were in the summer and autumn, which suggests that the cars were used to augment services in the holiday season but while this would have been needed on the Pembroke Dock branch, the Fishguard Harbour branch traffic was residential rather than seasonal.

Wolverhampton: The service to Stourbridge Junction listed on page N18 did not start until May 1925, when Car No. 40 was allocated, joined by No. 95 in November; both moved to Stourbridge in 6/26 when the service was worked from there. Nothing is known of the workings for which cars were stationed 5/13-11/15, or the three which were there 6/18-7/19, nor indeed the overlapping allocation of No. 54, 9/18-10/20, nor of No. 98, there 7-10/29.

Worcester: Following a gap 4/11-12/11, the allocation peaked at four for part of 1912/13 and there was another gap 1/20-7/22.

Yatton: The obvious duty for a car

stationed here was the branch to Clevedon, yet in 1911 (see Table 2A), this applied on Sundays only, the weekday duties covering a long trip to Swindon, out via Badminton and back via Chippenham, with a trip or trips to Avonmouth. At this time the Clevedon branch seems to have been worked by a Bath car. At some other unidentified date, the branch was railmotor-worked, based on a statement that a trip was made at 9.0 a.m. from Yatton to Wells to allow a freight trip to Clevedon. There were some gaps in the allocation down to 1917 with four cars at Yatton for part of 1909/10, reverting to two, with a gap from 3/17 to 1/19 with a single car being joined by another (described as shared with Bath!) from 3/20-1/21. There were two cars for part of 1923, one for part of 1925 and one (not shown under "period" in Table 2) from 12/27 to 8/29. After this the branch seems to have been worked by a Bristol car; there is photographic evidence thus on 20/3/35 (fig. P26).

Yeovil had a single car from 9/21 to 10/23 with an extra from 6/22 to 5/23 but no record of services worked has been found. The Yeovil Pen Mill to Yeovil Town is an obvious candidate, with possible workings along the line to Taunton as far as Martock.

It will be seen from the foregoing amplifications and amendments that the steam rail-motor story is complex (what is given above represents only the salient features) and the real answer is a separate book on the subject, which is beyond the scope of the present series.

TABLE 2A

SERVICES WORKED BY STEAM RAIL MOTOR CARS AND AUTO TRAINS IN THE SUMMER OF 1911

NOTES:

Saturday and Sunday workings which did not differ significantly from those on other weekdays are not shown separately.

Only the extremities of each working are shown except that where a unit spent a significant time (more than about half the total) on one sector within that area, this also is included, e.g. Stourbridge A - Halesowen Branch.

Unless otherwise shown the home station of the unit is the limit of travel in that particular direction.

Shed	*Limits of Working*
LONDON DIVISION	
Southall A SRM	Hayes, Westbourne Park, Gerrards Cross & Uxbridge, Kensington, Willesden
Southall B SRM	Westbourne Park, Gerrards Cross & Uxbridge, Willesden
Southall C SRM	Westbourne Park, Denham, Willesden
Southall D SRM	Brentford, Westbourne Park, Gerrards Cross & Uxbridge, Willesden
Southall Auto	Hayes, Paddington, Uxbridge (Vine St), Brentford branch services
Southall A (Suns) SRM	Paddington, Gerrards Cross & Uxbridge, Clapham Junction
Southall B (Suns) SRM	Paddington, Gerrards Cross & Uxbridge, Victoria
LONDON DIVISION — *continued.*	
Southall C (Suns) SRM	Paddington, Gerrards Cross & Uxbridge, Victoria
Southall D (Suns) SRM	West Ealing, Gerrards Cross & Uxbridge, Clapham Junction
Southall E (Suns) SRM	Paddington, Denham, Victoria
Southall F (Suns) SRM	Paddington, Greenford, Clapham Junction
Southall (Suns) Auto	Brentford branch services
Didcot SRM	Oxford, Reading, Basingstoke
Oxford A SRM	Heyford, Blenheim, Wheatley
Oxford B SRM	Princes Risborough, Kidlington

Shed	*Limits of Working*
LONDON DIVISION — *continued.*	
Aylesbury Auto	Princes Risborough, Banbury

BRISTOL DIVISION

Shed	*Limits of Working*
Trowbridge A SRM	Bath, Chippenham, Devizes, Patney, Frome, Warminster
Trowbridge B SRM	Bristol, Calne, Devizes, Patney, Westbury
Frome A SRM	Bridgwater, Taunton
Frome B SRM	Devizes, Hungerford, Warminster
Chippenham Auto	Calne, Bath, Trowbridge
Swindon SRM	Cirencester (Mondays)
Bath A SRM	Swindon, Chipping Sodbury, Yatton, Clevedon
Bath B SRM	Patchway, Nailsea, Chipping Sodbury, Trowbridge
Bristol A SRM	Keynsham, Avonmouth, Filton, Portishead
Bristol B SRM	Pilning, Portishead, Devizes, Trowbridge, Chippenham
Yatton SRM	Swindon, Avonmouth
Radstock SRM	Hallatrow, Westbury
Weymouth Auto	Abbotsbury, Dorchester
Weymouth SRM	Abbotsbury, Dorchester
Trowbridge A (Suns) SRM	Portishead, Salisbury
Trowbridge B (Suns) SRM	Bristol, Chippenham, Westbury
Bath A (Suns) SRM	Winterbourne, Charlton Mackrell
Bath B (Suns) SRM	WInterbourne, Avonmouth
Bristol A (Suns) SRM	Clifton Down branch
Bristol B (Suns) SRM	Clifton Down branch
Yatton (Suns) SRM	Clevedon branch
Weymouth (Suns) SRM	Abbotsbury, Dorchester

EXETER DIVISION

Shed	*Limits of Working*
Taunton SRM	Castle Cary, Bridgwater
Exeter SRM	Teignmouth services
Newton Abbot SRM	Kingswear, Brixham

Shed	*Limits of Working*
EXETER DIVISION — *continued.*	
Exeter (Suns) SRM	Newton Abbot services

PLYMOUTH DIVISION

Shed	*Limits of Working*
Plymouth A SRM	Wearde, Plympton, Tavistock
Plymouth B SRM	Wearde, Plympton
Plymouth C Auto	Saltash, Yealmpton, Plympton
Plymouth D Auto	Wearde, Plympton, Tavistock
Plymouth E SRM	Wearde, Plympton, Yelverton
Truro SRM	Newquay, Penzance
Newquay SRM	Camborne, Truro
Helston SRM	Truro, Penzance
Plymouth A (Suns) Auto	St. Germans, Plympton
Plymouth B (Suns) SRM	Saltash, Brent, Yelverton
Plymouth C (Suns) Auto	Wearde, Launceston, Yealmpton
Plymouth D (Suns) SRM	Saltash, Kingsbridge, Tavistock
Truro (Suns) Auto	Falmouth

GLOUCESTER DIVISION

Chalford & Stonehouse SRM services are not tabulated in the coach working programme although much reference is made to strengthening and ticket issuing by Conductors. Although a second car was allocated for two months in 1911 (and there were several other earlier and later cases) the services must have been maintained by one car which involved very intensive use.

Shed	*Limits of Working*
Cheltenham SRM	Honeybourne Moreton-in-Marsh
Forest of Dean branch	"Worked by Auto Engine & Trailer. A spare trailer is stationed at Bilson".

PONTYPOOL ROAD DIVISION

Shed	*Limits of Working*
Aberdare A SRM	Black Lion Crossing & Cwmaman Colliery
Aberdare B (Sats) SRM	Black Lion Crossing & Cwmaman Colliery
Aberdare C SRM	Mountain Ash, Swansea (East Dock)

Shed	*Limits of Working*

PONTYPOOL ROAD DIVISION — *continued.*

Shed	*Limits of Working*
Merthyr SRM	Llanwern, Newport
Monmouth SRM	Ross, Gloucester, Chepstow

SWANSEA DIVISION

Shed	*Limits of Working*
Neath SRM	Swansea (East Dock), Quakers Yard
Garnant SRM	Gwaun-Cae-Gurwen, Pontardulais
Aberayron SRM	Lampeter branch
Aberystwyth SRM	Lampeter
Whitland A SRM	Fishguard Harbour branch
Whitland B SRM	Pembroke Dock branch
Goodwick SRM	Fishguard Harbour, Whitland, Neyland

BIRMINGHAM DIVISION

Shed	*Limits of Working*
Stourbridge A SRM	Town, Dudley, Halesowen branch
Stourbridge B SRM	Town, Wolverhampton, Hagley, Halesowen
Stourbridge C SRM	Birmingham, Town branch
Stourbridge D SRM	Town, Langley Green & Oldbury branch, Smethwick Junction
Tyseley A Auto	Moor Street, Stratford, Bearley
Tyseley B SRM	Moor Street, Stratford, Bearley
Tyseley C SRM	Moor Street, Stratford, Bearley
Henley-in-Arden SRM	Lapworth branch
Stratford-on-Avon A SRM	Hatton, Moor Street, Henley & Lapworth branch
Stratford-on-Avon B SRM	Cheltenham, Evesham
Stratford-on-Avon C Auto	Leamington, Evesham
Banbury A Auto	Princes Risborough services
Banbury B Auto	Kingham branch

WORCESTER DIVISION

Shed	*Limits of Working*
Kidderminster A SRM	Stourbridge, Bewdley, Hartlebury
Kidderminster B SRM	Worcester, Bridgnorth, Woofferton

NORTHERN DIVISION

Shed	*Limits of Working*
Croes Newydd A SRM	Corwen, Oswestry
Croes Newydd C SRM	Ponkey branch, Coed Poeth, Moss
Croes Newydd D Auto	Berwig branch
Croes Newydd E SRM	Moss branch, Brymbo, Chester, Ruabon
Croes Newydd F SRM	Wynn Hall branch
Corwen SRM	Wrexham, Oswestry

GWR & LNWR JOINT

Shed	*Limits of Working*
Chester SRM	Birkenhead, Hooton & Helsby branch

ADDITIONAL INFORMATION AND CORRECTIONS RELATING TO PARTS 1 — 13

General Notes

In the Parts already published the words "Locomotive" and "Engine" have both been used but under a Society policy decision, "Locomotive" has been standardised. "Engine" will however be used when repeating information so headed in earlier Parts.

Additional information and corrections to earlier Parts are identified by the column and then the line in that column, on a given page. The line number is counted from the top of the page, including Chapter and paragraph headings, lines of dimensions and the like.

PART 1 — PRELIMINARY SURVEY

Page

12 — Column 2, line 6: The Lostwithiel & Fowey Railway, opened on 1st June 1869, was worked by the Cornwall Joint Committee as a branch of the Cornwall Railway and would not have needed a locomotive of its own.

17 — Table 1 (facing): Except for the T.V., the totals are of locomotives which were allotted Great Western running numbers; the T.V. total of 275 includes T.V. No. 333 which was not allotted a G.W. number. Some of these locomotives were withdrawn before they could be added to G.W. stock and the totals taken into stock should read 78, 34, 14 and 271 for the columns headed "Cam.", "Cardiff", N.B. and T.V. respectively. More details are given under the later "Part Ten" heading.

17 — Table 1 (facing): The official returns for 1922 are confusing, giving the total of Constituent Companies' stock in different places as 771, 797 and 798 as well as the Table 1 total of 799. Taff Vale 0-6-0 No. 333 was withdrawn without being given a G.W. number and the totals in the "1922" column of Table 1 of 76, 121 and 799 should be reduced by one to correspond to the correct official total of 798, the figures 42, 42 and 275 in the "T.V." column also being reduced to 41, 41 and 274 respectively. As detailed later under the "Part Ten" heading, 21 Cambrian, two Cardiff, one Neath & Brecon and three other T.V. 0-6-0's were withdrawn before they had been given G.W. number plates. Deduction of these 27 from the 798 total gives the alternative total of 771.
The official total corresponding to that of 861 in the last column of Table 1 is 859 and to agree, "2" under "G.V." in the Table should be altered to "1". The four-weekly returns show *Margaret* as taken into stock in 4.w.e. 28/1/23 and out of stock in 4.w.e. 25/3/23 but as the sale had been negotiated in 1922 (see page 220, Part Ten) the official version ignored the lady.

19 to 41 — Page 19, column 2, line 27 mentions the difficulty of discrepancies over locomotives recorded as built in December of one year and January of the next and, so far as it was possible the variant dates were shown in footnotes. A careful analysis of dates given in later Parts reveals a few other discrepancies and these arise because of differences between two official records. Page 14, column 2, line 25 refers to a new system of records, entitled ENGINE DATA which replaced the bound volumes of the 1870-1912 period, known as the REGISTER OF LOCOMOTIVE ENGINES. These volumes were not written up until the mid-1870's but the entries were back-dated to include post-1870 changes and there were less-detailed references to locomotives which had disappeared before the general stocktaking and renumbering which took effect from 28/7/70 (see page 9, column 1, line 12 onwards). It might have been expected that the dates in "Engine Data" would have been copied from the "Register" but another source must have been used as the dates often vary by a month and sometimes by more than a month.

A 1937 example will illustrate different versions arising through using the date of different events. The dates when new locomotives were taken into stock were generally a couple of days after they arrived ex Works at Swindon Running Shed, some up to four days later and, exceptionally, No. 4122 received from Works on 20/12/37 not taken into stock until 1/1/38. It seems likely that one set of records, kept by Works, would quote the actual day of completion and another the date taken into stock, the example of No. 4122 showing how this can bridge two years. Locomotives built by private firms were usually on trial for longer periods after arrival at Swindon, which would account for a difference between dates on makers' plates and dates into stock. Indeed, the same arose when it was Swindon and Wolverhampton practice to affix plates giving building dates. For example, the works plates on No. 111 *The Great Bear* were dated January 1908, this date was quoted in "Register" while "Data" gave February 1908 and the four-weekly returns show it taken into stock during 4.w.e. 29/2/08. "Register" gives the building date of the pioneer 4-6-0 No. 36 as August 1896, the works plates (Fig. M128, Part 12) show October 1896 and it was added to stock 4.w.e. 21/11/96.

Withdrawal dates are not relevant to Lot Numbers but it will be convenient to refer to discrepancies in the records in this regard which show the difficulty of vouching for a particular date, graphically illustrated by 2-4-0's Nos. 212/3, page C35, Part 3. These were sold to a Mr. Angus on dates variously given in the records as 9/12/1911, Dec. 1911, Jan. 1912 and Feb. 1912, while the annual returns show them still in stock as at 30/12/11. This is an exceptional case and perhaps Nos. 212/3 were incident-prone because, as recorded on page K79, Part 10, they managed to get wrongly recorded on Cambrian Railways' diagrams and on the G.W. Diagrams A3 and A4 which were derived from them!

Most of the discrepancies between records have been dealt within the relevant Parts but the following year-end overlaps need explaining, in so far as this is possible as those who compiled the Parts concerned are no longer with us and all that can be done at this stage is to quote the sources likely to have been used, showing "Register" as "R", "Engine Data" as "D" and a printed list of 1874 as "L":—

Part 1 page	Lot	Locomotive Number	Date	Source	Later page	Part No.	Date	Source	Notes
20	H	1103	1870	D & L	F17	6	12/1869	R	(a)
24	7	312/3	1864	D & L	D52	4	1/1865	R	
24	9	398/9	1867	D & L	D61	4	12/1866	R	
25	31	805	1873	D & L	D65	4	1/1874	R	(b)
30	167	3909 (c)	1908	—	E73	5	12/1907	D	(c)
31	172	3183-6	1907	(d)	J37	9	1/1908	D	(d)
31	187	4221	1913	D	J41	9	12/1912	R	
36	268	4981	1931	(e)	H32	8(2)	29/12/30	D	(e)

Notes: (a) — later No. 568; (b) No. 805 was added to stock in half year ending 31/12/73 and 1873 is preferable to 1874 in this context; (c) No. 3909 was converted from No. 2507 which was condemned in 1/08 but No. 3909 was fitted with its new boiler on 11/11/07, added to stock in 4.w.e. 4/1/08 and allocated to Birmingham on 6/1/08 which fits the "Engine Data" date of 12/07 better than the "Register" date of 11/07; (d) the footnote dates on Page J37 replace those in the Table; (e) See later reference to page 62, re No. 4981, also Nos. 5173/4, 5603, 6617/8 and 6749. Note also that footnote (iv) on page 23 gives the correct version and the dates against Lot N3 should read 2161-3, 1906, 2164-75, 1907, 2176-80, 1908.

Page 24 — Works Plates: Column 2, line 11 records that the last locomotives to carry works plates were those of Lot 183 of 1911, built at Swindon, while fixing of plates at Wolverhampton continued until the end of new construction there in 1908. There is very little earlier written evidence on the use of works plates and photographic evidence gets more scanty the earlier the date. What follows is distilled from the limited information available but as earlier photographs tended to be broadside views, the absence of a plate does not preclude its having been affixed at the front end; and as the footplate valance on some of the early locomotives was too narrow to accept a plate (the broad gauge "Prince" Class of

1846 is such a case, see Fig. B25 in Part 2) a plate may have been fixed on some other part of the framing, out of sight. Subject to these provisos, policy regarding affixing of plates and the information they carried seems to have been:

Broad gauge: There is photographic evidence that works plates carrying the building date (Swindon Works Numbers were not given to b.g. locomotives) were fixed to the first, the last and several locomotives in between of those built at Swindon and would be expected to have been fitted to all new construction even if not always visible on photographs. The standard plate was oval but a more elaborate square plate was used on the 4-2-2 shown at the Great Exhibition of 1851 (Part 2, page B20). Some at least of the "Standard Goods" 0-6-0's had two plates, one on each side of the centre wheel, the left-hand one recording building at Swindon and the right-hand one giving the date. However, the inevitable G.W.R. penchant for exceptions ensured that the "Sir Watkin" Class 0-6-0T had no plates; at least, none are visible on the photographs of four members of this small class of six.

Standard gauge, Swindon-built: Plates were carried on all locomotives from "1st. Lot Goods" (No. 57) down to and including Lot 183 but Works Numbers were not engraved on the plates until Lot 15, Nos. 445-454. A typical plate, fitted to 0-6-0ST No. 1286, appears in Fig. M127, Part 12. No. 3310 (later 4118) had an oval combined name, number and works cabside plate (fig. G42, Part 7); while still running as No. 3310, it acquired standard nameplates and the works plate was transferred to the usual position on the frames. Nos. 3332-95 (pre-1913 numbers) had a different style of oval plate in two varieties, as described on pages G20 and G33, Part 7. These carried the works numbers and building dates as shown by fig. G98 facing page G53, but even before renumbering took place, at least one of the building dates (on old No. 3379) had been ground off. Readers are referred to Part 7 for details of replaced nameplates and of renumbering, which are too complicated to summarise here but, using the new numbers, the dates are believed to have been removed from all but (new numbers) 3327/57, which retained them throughout (fig. M130, Part 12) and Nos. 3351/60 which still bore the dates on photographs dated August 1920 and August 1924 respectively. Figs. G96/9 facing page G53 and fig. M130 show plates with dates removed.

Standard gauge, Wolverhampton-built: The scarcity of photographic evidence has made it impossible to be precise, but the situation seems to have been that no plates were affixed up to and including Lot C (Works No. 70); they were fitted, including Works numbers, from Lot D to Lot L2 (Works Nos. 71-433). Sundry pictures of locomotives of Lots M2 to T2 (Works Nos. 434-511) do not show any plates and with Lots V2 to E3 (Works Nos. 512-624) the building date and Works number were cast on to the number plates, as shown on a plate from No. 2012 (Works No. 554), page M182, Part 12. As with the oval nameplates above, dates were often, but not always, removed, leaving the Works numbers intact. Photographs show no plates on locomotives built to Lots D3 to M3 (Works Nos. 625-764) and they seem to have been fixed to the frames, as was the case with No. 2128 derelict at Pontypool in 1939, this plate reading "Wolverhampton No. 732 June 1903". Finally the 2-6-2T built under Works Nos. 765-94 had oval plates below the smokebox door (Fig. J75, Part 9).

Works plates were specially revived for the L.M.S. Class 8F 2-8-0's built at Swindon in 1943-5. These were oval plates with four lines of letters and figures reading: LMS/BUILT 1943/ (or 1944 or 1945)/ G.W.R. Note that there were stops between the letters G.W.R. but not between L, M and S.

All British Standard locomotives and the "Quasi-standards" like Nos. 46503-27 (which were built at Swindon, as recorded elsewhere) carried similar oval plates but with three lines of letters and figures reading BUILT/date/SWINDON (or other factory where appropriate). Although 0-6-0PT Nos. 1650-69 were built three to four years after the first B.R. Standard, No. 75000, left Swindon, the Great Western tradition ensured that Nos. 1650-69 did not receive plates.

Page 26 to 41— A recently discovered source of information on the Works Lists and Swindon Lot Numbers makes it possible to fill in the gaps left in the Part One account and lists and to complete the story until the end of locomotive building at Swindon:

Lot 53 was issued in 1879 for renewals of ten of the "Sir Daniel" Class 2-2-2's (Page D9, Part 4) but was cancelled at an unrecorded date.

Lot 78 (Swindon Works No. 1158) was issued in May 1888 for "Engine No. 36, eight-wheeled saddle tank, 20" x 24" cylinders, 4' 6" coupled wheels" but no work seems to have been done and the order was cancelled in December 1893.

Lot 127 was issued to cover a further twenty "Metropolitan" 2-4-0T (page F28, Part 6) but was replaced by Lot 134 for twenty 2-4-2T which not only took the intended running numbers (3601-20) of Lot 127 but their Swindon Works Nos. 1866-85 as well (Page F38, Part 6).

Lot 230: These locomotives were ordered as Nos. 7320-9, see reference to page J13 on page M109, Part 12.

Lot 237: Kerr, Stuart & Co, received an order from the G.W.R. dated 14th November 1924 for overhaul and repair of Nos. 143, 238/46, 515/6, 619/25 and 938. Nos. 515/6 were cut up at the works in Stoke, while Nos. 619/25 and 938 were returned to Swindon. Lot 237 was either amended after the event or altered from the original order for eight overhauls. (See also page 33, Part 1).

Lot 266 originally covered 2-8-0T Nos. 5275-99, 6200-4 and Lot 270 was issued for a further ten to be numbered 6205-14. Owing to the depressed state of the South Wales coal industry, the orders for Nos. 5295-9, 6200-14 were cancelled with effect from 2/5/32.

Lot 308 originally covered Nos. 6800-99 of which Nos. 6880-99 were "deferred" in April 1941 but never reinstated. This alteration and the others down to Lot 353 were dictated by wartime priorities. Lots 318/9/20/31, originally for Nos. 7240-89, 3100-40, 8100-49 and 3220-39 respectively also were affected in the same way, terminating at Nos. 7253, 3104, 8109 and 3228 respectively. Work on Lot 324 was suspended after Nos. 5093-7 had been built, and the order for Nos. 5098/9, 7000-2 was cancelled with effect from 14/5/41. Lots 326/32 for Nos. 7003-7 and 7008-27 respectively also were cancelled from this date. On 20/5/41, Lot 340, originally for Nos. 6916-35, was extended to No. 6958, which seems to have been intended to compensate for the cancellation of the thirty "Castle" class, the mixed traffic "Hall" class being more suited to wartime requirements.

Lot 329 was originally for twenty 2-8-0T but Nos. 5265-74 were cancelled; a further twenty, to be Nos. 5275-94, were to have been built under Lot 343, issued 12/4/40 and cancelled on 14/5/41, apparently after work had started on the first seven.

Nos. 4140-9 of Lot 335 are a long way out of date sequence as the original order was cancelled on 14/5/41 and reinstated on 15/3/45.

Five Lots issued in the period for "8750" Class 0-6-0PT were all affected by wartime and immediate post-war conditions: Lot 336 for Nos. 4635-84 was cut back to Nos. 4635-60 on 20/5/41; a further 50, to be numbered 4685-99, 7600-34 were to have been built under Lot 344, issued 12/4/40 but cancelled 14/5/41; Lot 352 was issued 14/7/42 to follow the amended Lot 336 from No. 4661 to 4682, later increased to 4661-92, then re-issued to cover Nos. 4661-99, 9600-21 on 22/11/43; Lot 355 was originally for ten only and Lot 362 was originally for Nos. 9652-81.

Lot 339 for "2251" Class 0-6-0's Nos. 2231-50 was cancelled 14/5/41, only to be replaced by Lot 347 on 15/7/41, but over three years elapsed before No. 2231 appeared and completion of the order took nearly a year. The 0-6-0's on Lot 360 were originally to have been Nos. 8200-19.

Lot 341, for 60 and then 40 "2884" Class 2-8-0's for use in France has been described on pages 33 and N29 (Part 13). Nos. 3834-43 which *were* built under Lot 341 had been ordered under Lot 342, issued 12/4/40 and transferred to Lot 341 after cancellation of the original order. Perhaps some of the latter had been started, otherwise it would have been more logical to allocate the vacant Lot 341 to Nos. 5 & 6 (ex W.C.&P.) which were taken into stock under Lot 342.

Lot 345 (see page N29) was issued to cover preparation of 28XX 2-8-0's for service in France but was cancelled on 27/6/40.

The L.M.S. Class 8F 2-8-0's built under Lots 348/9 seem to have been intended to be numbered in G.W. stock when the orders were placed in February and April 1942 as numbers 3900-29 were allotted but they came out as L.M.S. Nos. 8400-29. Nos. 8430-9 were ordered as such under Lot 351. Lot 353 covered Nos. 8440-89 but the order for Nos. 8480-9 was cancelled on 1/2/45.

There were several changes of plan over the "County" Class 4-6-0's; Lot 354 was first issued for ten, to be numbered 9900-9 but increased to twenty, to be numbered 9900-19; Lot 358 covered Nos. 9920-34 when issued in 2/45; this was increased to 30 in 6/45 and reduced back to ten, 12/12/45. By the time Lot 359 was issued, they were referred to as the "1000" Class, but the fifteen covered by this order were cancelled. If all the plans had matured, there would have been 65 "Counties"!

The diesel shunters ordered under Lots 363/4 were originally intended to be numbered 501 and 502-7 respectively (see Part 11, second edition, page L24).

When Lot 373 was issued on 12/3/47, the numbers were to be 8400-9; on 8/10/47 these were altered to 3460-9; and on 16/3/48 to the final version, 1500-9.

The account on page 41 of Lots 382-9 was revised on page M66, Part 12 and, apart from Nos. 1650-69 built under Lot 417, no further Lot Numbers were issued for locomotives of "Great Western" design but as they are a part of Swindon Works history, a brief reference is made here. Full details of British Railways' standard steam locomotives and of Swindon-built L.M.S.-designed 2-6-0's will be given in other R.C.T.S. publications and the various types of diesel locomotive with a Swindon or Western Region connection have been amply covered in the commercial railway press. Some of the steam locomotives built at Swindon were allocated to other Regions and some of those built at other B.R. Works for the Western Region were given Swindon Lot Numbers (shown thus *).

Lots 390-3 were for Nos. 75000-9, 75010-9, 82000-9 and 82010-9 respectively.

Swindon had some influence on the design of L.M.S. Class "2" 2-6-0's Nos. 46503-27, built under Lot 394, as experiments with improved draughting had been made there on an earlier member of the Class, No. 46413. The tenders to Lot 394 were fitted with water pick-up gear, despite the intention of allocating most of Nos. 46503-27 to Mid Wales to replace the "Dean Goods" 0-6-0's! Automatic Train Control equipment also seemed something of a luxury on this Class.

Lot 395 recorded the transfer of 57 "Austerity" 2-8-0's to the Western Region while Lot 396 took ex-Shropshire & Montgomeryshire 0-6-0's Nos. 8108/82, 8236 and 0-4-2T No. 1 into stock for the sole purpose of withdrawing them, as none was serviceable. Lots 397-403 were for B.R. Standard locomotives, respectively Nos. 70015-24 *, 82020-9, 82030-4, 75020-9, 75030-49, 78000-9 * and 70025-9 *.

Lots 406/7 were 2-6-0's built for two other Regions with number sequences 77000-4/10-4 and 77005-9/15-9 respectively with Lots 408-10 covering Nos. 75050-64, 75065-79 and 82035-44 respectively. Lot 411 was originally for twenty 2-10-0's to be built at Crewe but only Nos. 92000-7 came to the Western Region.

Lots 413 (75080-9), 414 (77020-4), 415 (82055-62) and 416 (82045-54) were issued as the Modernisation Plan was being formulated and were deferred early in 1956, being cancelled later that year and none was built. Lot 418 covered the transfer of 119 locomotives from London Midland to Western Region stock in connection with boundary rationalisation.

Lots 421/2 covered construction of Nos. 92087-96 and 92178-92202 respectively, all for the Eastern Region while Lot 423 was for Nos. 73125-34 *. Lot 429 for Nos. 92203-20 included the last steam locomotive to be built at Swindon, and built for British Railways. This was No. 92220, a 2-10-0 appropriately named *Evening Star*, which received a copper-capped double chimney in the Swindon tradition and which was completed in March 1960 (Fig. M132, Part 12). It has been preserved in running order. The only other steam Lot was No. 433 for Nos. 92221-50, built at Crewe for the Western Region.

Diesel locomotives built at Swindon for the Western and other Regions or built by other Regions and by outside builders for the Western Region were covered by Lots 404/5, 412, 419/20, 424-8/30-2/4-59 and it may just be mentioned in passing that a shunter for Reading Stores acquired under Lot 431 for Service Vehicle Stock was numbered 20, which on the precedent of the petrol shunters described in Part 11, would have been the number the G.W.R. would have used if still in control of its destiny; and that the "Swindon touch" was perpetuated in the 200 H.P. Diesel-Mechanical shunters, which had vertical exhaust pipes reminiscent of a Great Western tapered copper-capped chimney. The last two batches of locomotives built at Swindon were ordered under a new series of Lot Numbers, British Railways Board Nos. 500/1 and the very last locomotive to be turned out from Swindon was a prosaic diesel-hydraulic 0-6-0 numbered 9555, which left the Works on 22nd October 1965, almost 120 years after the first left there in February 1846. This was the broad gauge 0-6-0 *Premier* which is shown in Fig. M131, Part 12.

Locomotive Liveries

Page

42 — Column 1, line 22: The model of *Iron Duke* is in the G.W.R. Museum, Swindon.

44— Column 1, line 57: Fig. P 27 shows the widely-spaced letters, in a much less classical style than usually used.

44 — Column 1, line 64: The "Great Western" lettering with "Garter" crest between also was used on locomotives fitted with prototype pannier tanks in 1910-2, see Figs. E79/84/5 in Part 5. No. 1047, in this case the second member of the "1016" Class to be so fitted, also had this form of lettering.

45 — Continued from foot of column 2: When the G.W.R. became part of British Railways on 1/1/48 it was expected that the Great Western tradition would have been submerged by a standard livery but after several experiments the final colour choice for most express passenger locomotives from the four Groups and of B.R. standard design was the G.W. Brunswick Green, although the lining-out differed from the G.W. style.

Details of individual class liveries appeared in Parts 4-11 and should be read in conjunction with these notes. Wartime shortages of labour and materials persisted into the early 1950's and affected plans to adopt a standard livery; men skilled in lining-out and lettering were not always available and lack of transfers of the Lion & Wheel insignia caused locomotives, including some new construction, to come out with blank tender or tank sides. Contrary to fears of regimentation, Western Region seemed able to interpret Railway Executive policy fairly flexibly and, in turn, edicts from Swindon to Wolverhampton and Caerphilly also were interpreted liberally rather than literally.

The first Railway Executive instruction with regard to liveries and numbering was that locomotives were to be lettered BRITISH RAILWAYS, the gap being left for a crest or emblem to be added later. Ex-G.W.R. locomotives were to have a "W" prefix added to the number, which in practice was painted underneath the number plate, and the first example so treated was No. 4946 which came out thus on the right hand side on 8th January, apparently for approval, and emerged from Works again on 16th, lettered on both sides. Four L.M.S. Class 5 4-6-0's were at an official inspection at Kensington Addison Road Station on 30th January in three shades of green (one of them G.W.R. green) and black, with various styles of lettering and lining. There was a further demonstration of repainted locomotives at Marylebone Station on 6th April. One example was Royal Blue, lined yellow, another light green, also lined yellow and a third black with full L.&N.W.-style lining. By 26th April the Royal Blue and light green examples had the lining changed to L.&N.W. style. A colour scheme, described as experimental, was published in May, providing that the most powerful passenger locomotives would be painted blue, other express passenger locomotives green and mixed traffic locomotives black, all with red, cream and grey lining. Freight locomotives would be unlined black. The Press Release was at pains to point out that the black would be the colour of "ripe blackberries".

In the meantime, tender and tank locomotives coming out of Swindon in January 1948 (and from Wolverhampton from February) mainly had BRITISH RAILWAYS in G.W. style letters but some came out unlettered.

The Western Region (W.R.) experiments comprised Nos. 6001/9/25/6 in ultramarine blue and Nos. 4089/91, 5010/21/3 and newly-built Nos. 7010-3 in apple green, all lettered BRITISH RAILWAYS in Gill Sans capitals. No. 6009 came out thus in May 1948, Nos. 4089/91, 5010/21/3 and 6001/25/6 in June and No. 7013 in July. The other six "Castles" in the current batch of ten, Nos. 7008/9/14-7 had G.W. standard green livery with BRITISH RAILWAYS in G.W.-style block letters. No. 6910 came out in the experimental lined black

EARLY DAYS, SUNDAYS AND LAST DAYS

Photograph] [K. Beddoes' Collection

Railmotor No. 35 at Stourport on inauguration of circular service Kidderminster-Bewdley-Stourport-Hartlebury-Kidderminster (shows either trials 29/12/04 or first service 2/1/05). P24

Photograph] [LCGB Ken Nunn Collection

Railmotor No. 89 on Sundays-only service leaving Paddington for Greenford 21/8/10. P25

Photograph] [J. G. Dewing

Railmotor No. 92 entering Clevedon 2/3/35 (All steam railmotor services ceased in the Autumn of 1935). P26

A SELECTION OF CURIOSITIES

Photograph] [Peter Treloar's Collection

No. 3213 at Oxford in 1904 with tender lettered in a much less elaborate style than generally used. P27

Drawing from "Engineer" of 3rd January 1896 showing *Ajax* and/or *Mars* with streamlined front end (Compare Fig. B5 in Part Two). P28

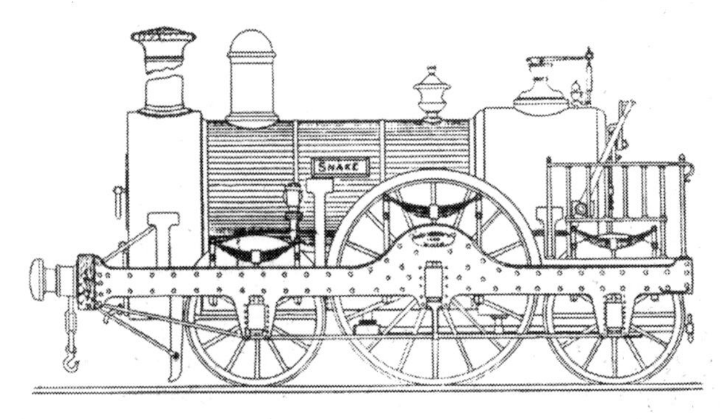

This drawing from "Railway Club Journal" July 1903 seems to show *Snake* and *Viper* as rebuilt in 1839/40. P29

"KING" CLASS IN TWO BLUE LIVERIES

Photograph] [K. H. Leech/Colour-Rail

No. 6025 *King Henry III* **in experimental ultra-marine blue livery and tender lettered BRITISH RAILWAYS, leaving Chippenham with an up train in 1950.** P30

Photograph] [P. M. Alexander/Colour-Rail

No. 6004 *King George III* **in standard light blue livery leaving Box with an up stopping train in 1952.** P31

livery with BRITISH RAILWAYS in Gill Sans letters on its tender in June 1948. The lettering was painted out and Lion & Wheel emblems applied in July; Nos. 5023 and 6009 were dealt with in the same way at the same time. No. 5954 came out in lined black at the end of July but in this case with the Lion & Wheel emblem on its tender.

The Railway Executive decided upon a lighter shade of blue for the most powerful passenger locomotives and the whole of Nos. 6000-29 were so repainted, starting with No. 6000 in June 1949. They chose dark green for the other express passenger locomotives, which was applied to the "Castles" and the surviving "Star" class 4-6-0's. Blue locomotives were lined out in black and white and green locomotives in black and orange. Nos. 6000-29 exchanged their blue livery for dark green between April 1952 and February 1954. The two shades of blue and the apple green liveries are reproduced in Figs. P 30, 31 and 32.

From July 1948 there were isolated cases of lined black livery but most locomotives were in unlined G.W. green, the last new from shops being No. 7438 (to stock 6/10) and 4165 (to stock 8/10). W.R. had announced that lined black would be applied to all two-cylinder 4-6-0's except 29XX, to all 4-4-0's, 0-6-0's, 2-6-0's, 0-4-2T, 2-4-0T, Taff Vale "A" Class and B.&M. 0-6-2T and all G.W. 2-6-2T, to 54XX (but not 64XX) and to M.S.W.J. Nos. 1334-6. Freight (including Nos. 4700-8) and shunting locomotives to be unlined black. The 29XX were added to the lined black category early in 1949 and there is no record of any being turned out in lined green. Nos. 2920/6/7/34/7/45/7/9/54 were repainted in lined black as from February 1949 (Fig. P33). The remainder of the class either did not survive until 2/49, or are known not to have been repainted between that date and withdrawal. Photographs and personal observations show that apart from 4-6-0's, lined black livery seems to have been confined to 4-4-0's Nos. 9009/14, 0-6-0 No. 2579 (Fig. M33, Part 12), which was used in comparative trials with B.R. 2-6-0 No. 46413 (see page P41), 0-6-0 No. 2238, 2-6-0's Nos. 7313, 9314/9, a few 0-4-2T, 2-6-2T Nos. 4406/9, some 45/55XX, 41/51XX (but not 31, 61 or 81XX) and 0-6-0PT No. 5409. Owing to the shortages mentioned, while Nos. 4166/7 came out (in October 1948) in lined black, Nos. 4168/9 were unlined and Part 9, page J31, column 2, line 50 should be corrected. The inevitable exceptions to standard instructions also gave No. 4702 lined black livery and Nos. 1503-5, 8762-4/71/3 were similarly painted, no doubt in connection with their empty carriage workings to Paddington station.

The B.R. "Lion & Wheel" emblem replaced the Gill Sans lettering, first seen on new construction on No. 7018 (to stock 24/5/49) but shortage of transfers again led to some locomotives coming out with blank tender or tank sides.

Oddments included red backgrounds to the number plates of some of the smaller 0-6-0PT overhauled at Crewe and Derby in 1949-51 while in 1950-53 some 4-6-0's had red backgrounds to both name and number plates. On the Vale of Rheidol section, 2-6-2T No. 7 was reported repainted unlined green in April 1955 and Nos. 8 & 9 were similarly repainted in 6/56, when all three received names. These changes would have been related to the line's tourist status and do not suggest a change of policy. For later livery changes on Nos. 7, 8 & 9 see page N5, Part 13.

Although Swindon gave instructions at the end of 1956 that all 4-6-0's were to be in lined green and all other passenger and mixed traffic locomotives in unlined green, this seems to have been anticipated informally as No. 1020 came out in lined green in May 1955. The "Counties" and "Modified Halls" were chosen for this treatment and by the spring of 1956 Nos. 1000/12/26, 6972/90/4/5/7 and 7902/4 had been reported. Of the earlier "Halls" Nos. 4981 and 5910 came out thus in June 1956 and by the autumn, some "Granges" and "Manors" were similarly repainted. 2-6-0's Nos. 6372/85 were repainted as a special case in May 1956 for working the Royal Train over a route barred to heavier locomotives.

In late January 1957 the instruction was revised to require lining on all passenger and mixed traffic locomotives, which usually was accompanied by polishing of safety-valve covers. From March 1957 the second style of B.R. crest (Fig. P35) replaced the "Lion & Wheel" and seemed more in keeping with the new livery.

This was not just a welcome return to widespread use of G.W. green but an advance in the quality of the livery as lining out had been confined to express passenger locomotives in G.W. days. Swindon, Wolverhampton and Caerphilly factories interpreted the instruction differently according to the duties of various classes in their sphere of influence. Caerphilly turned out at least 126 of the 200 56/66XX 0-6-2T in this livery on the grounds of their wide use on Cardiff Valleys passenger trains. The figures thus (19) are the totals noted in each case and the actual numbers are probably higher: 14XX (19), 22/32XX (34), 3100-4 (3103 only), 41/51/61/81XX (100), 43XX (4358 only), 45/55XX (56), 5111-49 (5148 only), 53/63XX (52), 54XX (5), 64XX (21), 7300-21 (12), 7322-41 (14). Because 2-8-0's Nos. 4700-8 were classified 7F and regarded as freight locomotives, they had been painted plain black but commencing in mid-1957 they were repainted in lined green (Fig. P34) and in 1958 these impressive machines were deservedly reclassified 7MT (but see No. 4702 on page P43).

The combined effect of dieselisation of local passenger services and closure of branch lines was to relegate steam locomotives to freight duties and it was not surprising to find that the first examples of unlined green 56/66XX were noted in February 1961, since which date lined green livery disappeared, either by repainting or withdrawal. In some cases, paintwork was neglected and later coats weathered away to reveal earlier liveries, the most striking example being No. 8762 which had been repainted lined black (see above) but by 29/10/62 this had worn away to reveal the letters GWR. These letters also were visible on No. 4628 when withdrawn in May 1964.

The original B.R. livery for the diesel railcars described in Part 11 was crimson and cream, except for the parcels cars Nos. 17 & 34, which were painted crimson overall. The cars which survived into the last phase of repainting came out in green overall, more than one shade being used at different times. The auto trailers described in Part 11 were sometimes painted crimson and cream in the early days but crimson overall was more common, especially towards the end of auto working.

Page
46/47— **Engine Diagrams** (as amended by Page M67, Part 12): there are some gaps in the alphabetical/numerical sequences of Diagram Indices and some cases of duplication. There are no Diagrams for 0-6-0 W, X and A22, or for 0-6-0T B25 and B32, neither is there any trace of Diagram 0-6-0T B29 although the alteration to No. 818 was carried out as described on page K255, Column 2, lines 32-36. Diagram 2-8-0 N was first prepared, perhaps only in draft, for 2884 Class 2-8-0's with a larger tender for overseas wartime service and was re-allocated to L.N.E.R. 2-8-0's on loan to the G.W.R. in 1940-3 (see page M41) but this does not seem to have been issued, probably because the three quite significant variations in the L.N.E.R. locomotives could not have been shown on one drawing. 0-4-0T Diagram K was first issued for *Gallo* (see page C96) and used later for No. 1340 (see page K16).

Diagrams 0-6-0T B58 and 0-6-2T A23, A29 and A45 were prepared respectively for alterations to Nos. 604-6 (page K122), 154 (page K94), 1084 (page K207) and for fitting of Standard No.3 boilers to T.V. "A" Class (page K189). No Diagram was issued for 4-4-2T No. 2230 with Standard No. 4 boiler (page J26) nor for Nos. 250/68 or 410/37 with G.W. cabs, tanks and bunkers (pages K31 and K184).

Steam Railmotor Diagrams were numbered A, A1, B-D, D1, E-G, G1, H, J, J1, K, K1, L, M, M1, N-Q, Q1, R-T, see "Revised Table 2" on page M127.

47 — Add to Column 2 after line 15: New Diagrams were not issued to record new boiler variations in later years, the Diagrams being altered without variation of the index letter and number. This is known to have applied to 2-6-0, 4-6-0, 2-8-0 and 2-8-2T Diagrams, Sometimes the original figures have been crossed out and sometimes completely erased. This will explain the lack of separate Diagrams for 68XX 4-6-0's with AK boiler and 72XX 2-8-2T with DL boiler, referred to on page M67 against "Page 47 — Column 2".

48 to
58 — **G.W.R.-built boilers:** Despite the initial extra cost of copper inner fireboxes compared with steel, they were far superior in life expectation but the G.W.R. had an open mind and experimented with a steel firebox on a Standard No. 1 boiler (No. 2873) which was put on 2-8-0 No. 2835 from May 1913 to November

"CASTLE" AND "SAINT" CLASS EXPERIMENTAL LIVERIES

Photograph] [K. H. Leech/Colour-Rail

No. 4091 *Dudley Castle* **in experimental apple-green livery and smokebox number plate, with tender lettered BRITISH RAILWAYS, at Chippenham, 1949.**

P32

Photograph] [T. B. Owen/Colour-Rail

No. 2934 *Butleigh Court* **in lined black livery with red-backed name and number plates and "Lion & Wheel" emblem on tender, ex-Works at Swindon Running Shed, June 1950.**

P33

LATER STANDARD LINED GREEN LIVERY

Photograph] [P. B. Whitehouse/Colour Rail

No. 4704 in lined green livery with the second style of B.R. crest, at Shrewsbury Shed. P34

Photograph] [D. T. Rowe

No. 5571 in lined green livery with the second style of B.R. crest (note that back of bunker is unlined) at Pembroke Dock Shed, July 1962. P35

1914, was then used on Nos. 2983/19/38 and was fitted with a copper firebox in March 1922, lasting thus until September 1931.

Later examples seem to have been dictated by wartime copper shortages and twenty steel fireboxes were fitted to "2301" Class boilers in 1918/9, lasting for periods extending from 1923 to 1928. Ten were fitted to "517" Class boilers in 1919/20, lasting from 1924 to 1931. Fifteen "Sir Daniel" Class boilers were fitted with steel tubeplates only in 1918, with a life of from five to eleven years.

Only four steel fireboxes were fitted during the Second World War, all to Standard No. 1 boilers built in 1944. Two of rivetted construction were put on Nos. 5917 and 6870 and two of welded construction on Nos. 5988 and 6967.

Oil Firing: After the end of World War 2, there was an acute shortage of good locomotive coal but a better supply of fuel oil and a programme of conversion to oil firing was put in hand in 1945, with the intention of converting 85 "Halls", 25 "Castles", 73 2-8-0's and a single 2-6-0. The programme ran into difficulty, not with supplies but with distribution and only eleven "Halls", five "Castles", twenty 2-8-0's and the solitary 2-6-0 (No. 6320) had been altered when the programme was deferred in September 1947, never to be revived. Individual details were given in Part 8 (1) pages H16 and H30, Part 8 (2) pages H16 and H31 and in Part 9, pages J14 and J21.

Page

61 — Column 1, line 35: In the early stages of the design of the device for automatic clipping-up of the A.T.C. shoe, an experimental unsprung shoe was fitted to a locomotive, believed to have been No. 4037 and, following further work, a standard version was fitted to "Hall" Class 4-6-0's Nos. 5931-40, to 2-6-2T Nos. 6100-69 and 0-6-0PT Nos. 9700-10. They were able to work into Paddington suburban station and to Addison Road. From time to time the latter station handled special excursion trains and the "Halls" would have been used on these and on commuter trains starting from Paddington suburban station. Nos. 9700-10 could also work to Smithfield Market.

61 — Foot of column 2: British Railways standardised an alternative system known as A.W.S. and the Great Western system was gradually superseded. In the meantime, most, if not all, of the larger, and some at least of the smaller, locomotives on loan or transferred from other Regions of British Railways were fitted with the Great Western system. British Standard locomotives built at Swindon were turned out with the equipment, which was fitted on arrival at Swindon to Standard locomotives built at other factories for Western Region use. It is likely, but not certain, that the fairly large scale transfers from other Regions following rationalisation of boundaries also were so fitted on arrival, although, as mentioned elsewhere, not all the Swindon-built L.M.S. Class 8F 2-8-0's received A.T.C. equipment when they returned to ex-Great Western metals in 1954-6.

62 — **Four-weekly periods:** Research by Peter Jones and Jim Oldham has revealed a four-weekly period ending on 24th January 1846, with the implication that the statistical year ended on 27/12/1845 rather than 31/12/45. This suggests that the system could have been in use earlier. Periods of two weeks only also were used. Using fixed-length periods suited the accountants but made the lot of the historian harder as the nearer the last day of the accountants' year crept towards 3rd December, the more locomotives built in that year were credited (by the accountants) to the following year. The men who had to use the locomotives preferred the actual date of entry into service and there are only a few recorded instances of dates being qualified. These relate to Nos. 5603, 6617 & 6618 (Part 5, page E76), No. 6749 (Part 5, page E81), No. 4981 (shown as 12/30 on page H32, Part 8, 2nd Edition, should have been qualified by a footnote (h) on page H35, 2nd Edition, Part 8, reading "Taken into stock in 1931 accounts" to the 1931 date against Nos. 4981-99 on page 36, Lot No. 268). The only others were Nos. 5173/4 (Part 9, pages J32/3).

The exact significance of the dates quoted in the different Swindon records is not clear. For example, although the "Register" shows "City" Class 4-4-0's Nos. 3436/7 (later 3713/4) as built in March 1903, neither was weighed until 23/4/03, "Engine Data" dates both May 1903, they were added to stock in 4.w.e. 2/5/03 but had been sent out to their first sheds on 27/4/03 and 29/4/03 respectively. The final stage in building a locomotive is to weigh it and to make any adjustments to weight distribution, whereupon it is ready to hand over to the Traffic Department and the effective date for Nos. 3436/7 would have been April 1903. Other differences between written records and between them and dates on Works plates are recorded elsewhere and it was not until 1926/7 that "Engine Data" recorded the exact day of handing a locomotive over to the Traffic people. Nos. 3020-99 were the first to be so treated, starting with No. 3051 dated 11/1/26, followed by No. 5667 (dated 16/6/26) and on to No. 5699. There was a gap until the practice became general, starting with No. 5006, dated 3/6/27, after which exact dates were quoted for all new construction.

The same system continued under British Railways, Western Region but they managed to make the totals at the end of the last four-weekly period agree with the 31st December totals, either by not making any alterations in the interim or by adding an extra week to the last period. (The only exception was not a "G.W." matter, Standard 2-6-0 No. 78003 completed 31/12/52 appeared in 4.w.e. 24/1/53 and in the 1953 totals). No. 3409, the last locomotive to be built to a G.W. design, was included in the return for 4.w.e. 3/11/56 and 2-10-0 No. 92220, the last steam locomotive built at Swindon, was shown in the return for 4.w.e. 26/3/60. Although there was steam working up to 3/1/66, all remaining W.R. steam locomotives were taken out of stock in 4.w.e. 1/1/66. The very last locomotives of G.W. design, Nos. 4646/96 and 9774, (except for the Vale of Rheidol locomotives) were all taken out of London Midland Region stock on 12/11/66, in their Period 12/66 ending on 3/12/66.

PART 2 — BROAD GAUGE

Page

B10 — Column 1, line 47: A detailed account of the building of *Ajax* and *Mars* by a man who had worked on them, together with a drawing (fig. P28) appeared in the *Engineer* of 3rd January 1896. An extract from the account reads "The engine was designed by John Grantham, draughtsman at Mather, Dixon & Co., North Foundry, Liverpool. The outside view resembled a steamer; the driving-wheel splashers like a paddle-box, and the handrail plates, brought to the buffer planks, shaped like the stem of a vessel, and intended to take the wind pressure off the front of the engine". If the drawing and the description are correct, this would have been the first attempt at aerodynamic design and they also are referred to in an American publication "The Streamline Era" by R. C. Reed but Gooch made no mention of this feature in his Diaries and in evidence to the Gauge Commissioners in 1845, Brunel referred to the adverse effect of *side* winds on the 10ft. diameter plate wheels, with no mention of the front end. However, for what it may be worth, the wheelbase given as "about 15ft." on page B10 scales 15' 10" on drawing B5 and 15' 4" on drawing P28 which lends a shred of authenticity to the latter.

B11 — Column 1, line 30: More details are to hand about *Snake* and *Viper*, some of it contradictory. G. A. Sekon's "Evolution of the Steam Locomotive" (1899) quotes a statement from Gooch's Diaries that the gearing was 2:1, not 3:2 and also quotes his opinion that they would be unlikely to do more than propel themselves; yet Sekon also quotes an eye-witness of August 1838 who described a journey with five carriages from Paddington to Maidenhead at an average of 36 m.p.h. !

Even so, they were altered to more conventional locomotives in 1839-40 (see page B11) and although the drawing reproduced as Fig. P29 from the "Railway Club Journal" of July 1903 is described as showing them as built, the drawing surely must show the 1839-40 alterations as there is no sign of gearing. The block from which Fig. P29 is reproduced was made for Mr. Melling, at one time manager of the Haigh Foundry, "from the original signed working drawing in his possession" which seems to imply that the makers made the 1839-40 alterations.

The 1903 account says that they were "numbered 13 and 14 in the G.W.R. locomotive list", the only known reference to numbering of broad gauge stock and presumably order numbers, not running numbers. Apart from the cylinders, dimensions differ from those given on page B11 as follows: driving wheels 6' 0",

TWO VERY DIFFERENT COLOUR SCHEMES

Photograph] [Welsh Industrial & Maritime Museum

No. 450, preserved and restored as Taff Vale Railway No. 28 in T.V. lined-out black livery (see Part 10, Page K137). Photograph taken at Caerphilly. P36

Photograph] [P. Zabek

London Transport No. L90 (formerly G.W. No. 7760) in L.T. Maroon livery, a snow scene at Watford Tip, January 1968. P37 carrying wheels 4' 0", boiler barrel 8' 0" long, containing 111 tubes 1.5/8" diameter, total heating surface 500 sq. ft. The original wheelbase was 10' 6" and there must have been a major alteration to the frames to accommodate the 13' 0" wheelbase after rebuilding shown on page B11.

Page

B11 — Column 1, line 56: Sekon's book also includes line drawings of *Hurricane* and *Thunderer.* These are reproduced in the text without "Fig." numbers as there can be no guarantee of their authenticity but that of *Thunderer* is certainly more credible than Fig. B8 in Part 2. Phil Reed, compiler of Part 2, was a stickler for accuracy. He must have been aware of the existence of these drawings and may have had good reason for not including them and they are reproduced here with this cautionary note.

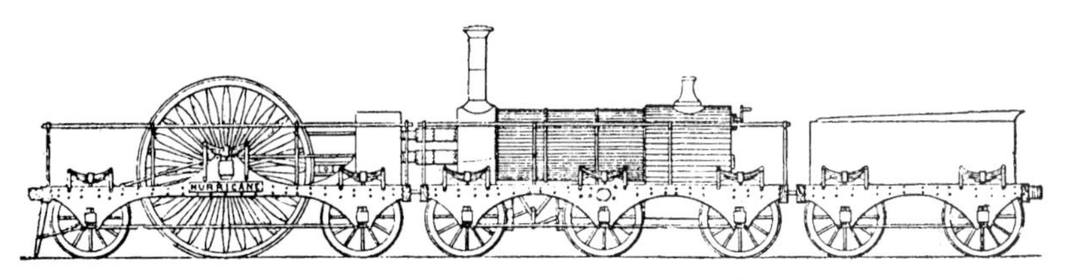

THE "HURRICANE", A BROAD GAUGE ENGINE WITH 10ft. DRIVING WHEELS, BUILT ON HARRISON'S SYSTEM

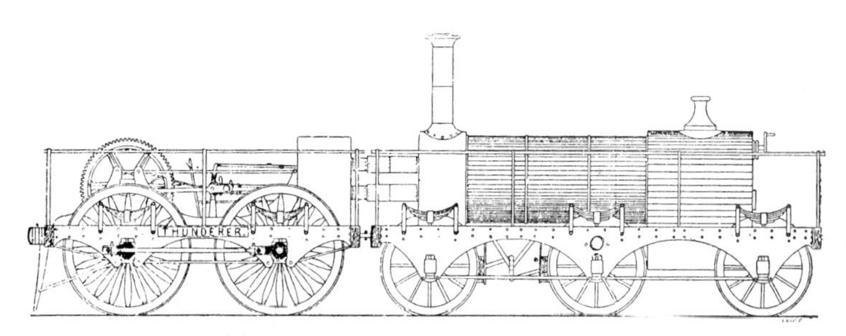

THE "THUNDERER," A BROAD-GAUGE ENGINE BUILT ON HARRISON'S PLAN, WITH DRIVING WHEELS 6ft. DIAMETER, GEARED UP TO 18ft.

B25 — Column 2, "9th Lot Goods". The name *Chronus* should read *Chronos.*

B36 — Column 2, line 40, also supplementing information on page M70: Rowland Brotherhood had contracts for railway construction for the Bristol & Gloucester Railway and for the G.W.R. In his diary in 1858 he refers to Peter Brotherhood building a broad gauge locomotive *Moloch* in the old shed outside his works. In 1863 a 6-coupled broad gauge locomotive with 4' 8" wheels and 17" x 24" cylinders was offered for sale but there is no record of any buyer. Rowland Brotherhood of Chippenham had a forced sale on 17th November 1869 when locomotives offered included a broad gauge 6-wheel-coupled tank engine with 16½" cylinders; again no sale was recorded. The temptation to link *Moloch* with B.&E. No. 111, on the grounds of marked similarities, must be tempered by the reference to a "tender engine" in a G.W. list (see page M70).

B50 — The end of the Broad Gauge was chronicled here in prose and in pictures facing page B55 and it is apt to end Part 14 with a poetic version (see page P158).

B56 — Information on the later history of some of the sold broad gauge locomotives has come to light, as follows:

Alexander worked at the Avon Colliery, Blaengwynfi which had been acquired by the G.W.R. in 1874 and the sale was a paper transaction. *Alexander* ceased work at the colliery in 1907, it was sold to Robert Gibb in 1910 and was used as a haulage engine at the nearby Glengarw Colliery for many years.

Argus was used as a winding engine, probably at Nailsea Colliery.

Flora was advertised for sale by C. D. Phillips of Newport Mon. (now Gwent) between 7/1/81 and 2/4/83., described as a "Broad gauge loco., used for pumping, 17" cylinders, £400".

PART 3 — ABSORBED ENGINES 1854 - 1921

Page

C26 — Column 1, line 19: The part-frames, wheels and motion of No. 252 were still at Wolverhampton in September 1961. The works closed in 1964 and, probably at that time the "252" exhibit was transferred to the then Staffordshire County Council Industrial Museum at Great Haywood, Stafford. The exhibit later was transferred to Shugborough Hall, the new Staffordshire County Museum 5 miles S.E. of Stafford.

C27 — Column 1, line 26: A locomotive *Viso* appears in an Ebbw Vale list dated 1874.

C53 — Column 1, line 60: The Ebbw Vale Steel, Iron & Coal Co. Ltd. had an outside cylinder 0-6-0ST named *King Edward VII* which bore a plate saying that it was rebuilt at Ebbw Vale Works in 1902, the year in which Edward VII was crowned. It worked later at John Vipond & Co.'s premises, Varteg, where Fig. P38 was taken. Below the running plate there is a marked resemblance to the 0-6-0 shown in Fig. C109, facing page C53 which suggests that it was M.R.&C. No. 5. Line 40 records the sale of a similar 0-6-0 to the Coalbrook Vale Colliery, which also became an 0-6-0ST and it is just possible that the conversion was carried out at Ebbw Vale (only two miles from Coalbrookvale) but this locomotive was registered for working over the G.W.R. near Nantyglo (No. 14) in 1900 (plate presumably fitted May 1901). The registration numbers were re-used promptly when a locomotive registration lapsed but the number 14 was not used again until 1914, so that conjecture that the Coalbrook Vale locomotive is the subject of fig. P38 does not seem tenable. (See pages N20 and N22). The hiring referred to in line 41 appears to have started about 1872 as the G.W. Registration Book has a note to this effect. The withdrawal date of the former M.R.&C. No. 4 is not recorded but the colliery itself does not appear in a list of South Wales collieries dated 1917 and, of course, the registration had expired by 1914.

C65 — Column 1, line 14: C. D. Phillips, Locomotive Dealer of Newport, Mon. (now Gwent) and Gloucester advertised from October 1901 to June 1902 a "3' 0" gauge locomotive by G.W.R., outside cylinders, fitted with Gooch's outside link motion gear, 5' 0" wheelbase, tubular Cornish boiler suitable for a w.p. of 120 lbs, water tank fitted between the frame. The whole in very good working order throughout, approx weight 6½ tons." Described in advertisement dated 1/10/01 as "now on hire in Gloucester". This is clearly one of Nos. 1381/2 and although some of the dimensions do not make sense, it seems likely that the 3' 0" gauge locomotive advertised by the Bute Works Supply Co. of Cardiff on 27/7/00 and 11/1/01 is also one of these. Rivers Department archives at Davyhulme clearly describe their locomotive as G.W. No. 1382 but do not record the date of acquisition and it is not possible at this stage to identify which of the two locomotives was involved in the various advertisements.

C67 — Column 1, lines 1-3: The locomotive was first reported in Purfleet in 1912 and a report of a 1917 sighting says it still bore the number "2". *Locomotive News & Railway Notes* for 10th January 1920 said "ex G.W. No. 2 is at present working in a margarine factory at Purfleet", which the Industrial Locomotive Society says probably was the Stork Margarine factory. This Society's Anglo-American list has a pencil note "Elsie - 2-4-0T". The advertisement mentioned in line 6 described it as a 2-4-0 saddle tank with 10" x 18" cylinders, 3' 0" coupled wheels, 4' 9" coupled wheelbase, 125 lb. pressure.

C69 — Column 1, line 57: Although a large number of Neath Abbey Ironworks drawings still exist, none fits the description of *Ironsides*.

TWO PROBLEM PICTURES AND A GAUGE CONVERSION

Photograph] [F. Jones' Collection

King Edward VII **at Vipond's, Varteg, at a date after 1902. Probably a rebuild of M.R.&C. No. 5, Fig. C109, Part 3. (The original photograph was damaged).** P38

Photograph] [F. Jones' Collection

Duty **at Abergorki Colliery is very much like Fig. C167, Part 3, apart from the cab.** P39

Photograph] [David Tipper's Collection

S. & W. *Robin Hood* **after conversion to standard gauge.** P40

BROAD AND NARROW GAUGE TANK LOCOMOTIVE—SEVERN AND WYE RAILWAY.

BY MESSRS. FLETCHER, JENNINGS, AND CO., ENGINEERS, LOWCA WORKS, WHITEHAVEN

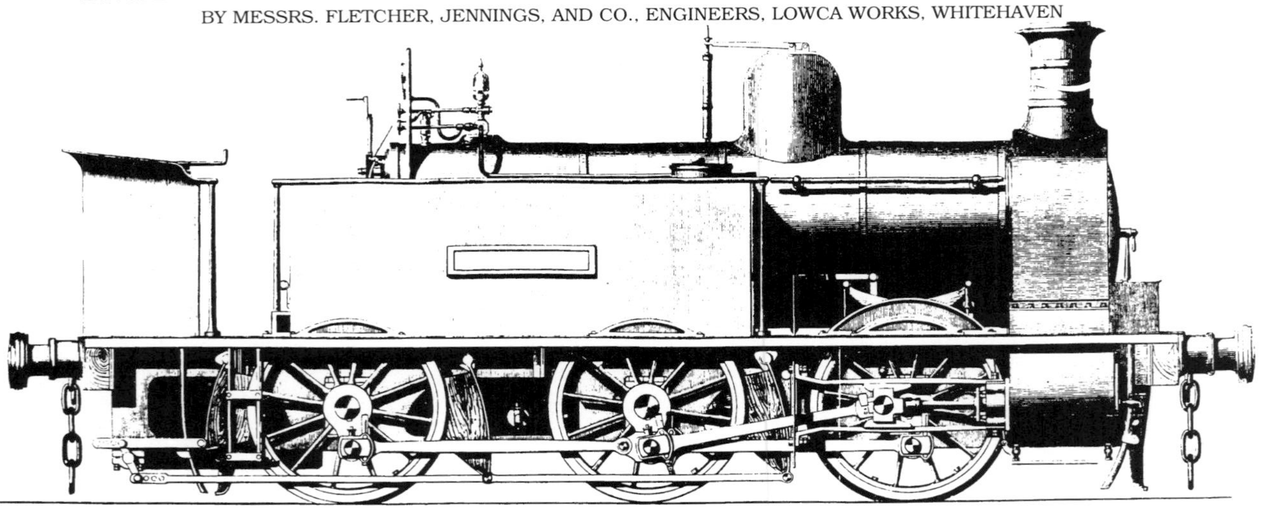

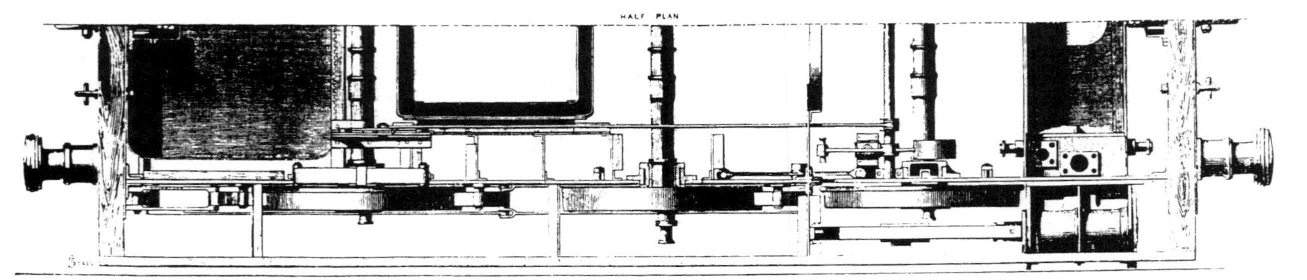

[Reproduced from Maker's Drawing

Half-Plan and Side Elevation of Severn & Wye *Robin Hood* as built for the broad gauge with provision for conversion to standard gauge.

P41

C72/3 - The last line in the caption to line drawing No. C148 should read "believed to be W.C.R. *Carn Brea*".

C74 — Column 1, line 21: Although a photograph of ex-G.W. No. 1398 at Sharpness Docks does not show any number, it was No. 2 in the Sharpness stock list. A new locomotive supplied in 1902 bore number plates "S.D. No. 3". Another secondhand locomotive acquired in 1920 was known informally as "No. 4" but carried no plates.

C74 — Column 2, line 17: No. 10 was sold by Sharp, Stewart & Co. to the C.V.&H.R. in 1880.

C80 — Column 1, line 12: In January 1926 it was decided to rebuild No. 1387 to replace No. 1337 (Part 3, page C91) on the Weymouth Quay line, No. 1387 in turn to be replaced at Reading Signal Department by Simplex Petrol locomotive No. 27 (Part 11/2, page L22). In the event, No. 1387 (renumbered 1331) was not returned to service until 26/11/27, although No. 1331 had been condemned on 16/1/26. As a Departmental locomotive, No. 1387 was not in running stock but was allocated the number 1331 and shown as an addition to stock in four weeks ending 20/2/26. For no obvious reason and while it was still in Works, it was taken out of stock in 4.w.e. 20/3/26 and reinstated in stock in 4.w.e. 17/4/26.

C81 — Column 1, line 6: recent research reveals both confirmation and confusion over the history of No. 2. It was wrongly described as *Dusty* on page M72, Part 12; the name was *Duty* and the photograph (Fig. P39), presuming that it had been B.P.R.&P. No. 2, must have been taken in or after 1890. The Shropshire & Montgomeryshire Light Railway (which had earlier operated under other names) was reopened as such on 13th April 1911, and the locomotive which became No. 2, *Hecate* was reported as purchased from the Griff Colliery, Nuneaton in May 1911 "through Mr. Hartley" (who may have been a dealer). But another account says that the same locomotive was inspected at Barry by a S.&M. fitter with the implication that it had remained in South Wales until 1911. S.&M. records show that it was condemned in 1931 and was noted outside Kinnerley shed, partly dismantled, in 1932 and 1933, being finally cut up in April 1937. It had been renamed *Severn* in 1921. To add to the general confusion, there were two Abergorki Collieries, at Mountain Ash and at Treorchy and although the photograph is captioned "Mountain Ash", Messrs Burnyeat Brown's colliery was at Treorchy.

C83 — Column 1, line 57 (and page K262, column 1, line 27): An 0-4-0ST which almost certainly had been Powlesland & Mason's No. 8 or No. 9 was seen at Birtley, six miles south of Newcastle-on-Tyne in 1915.

C84 — Column 1, line 29: Fig. P41 reproduced from *The Engineer* for 30th April 1869 shows *Robin Hood* as built as a broad gauge convertible locomotive. A water tank under the footplate had to be removed but otherwise the conversion mainly involved placing the main frames, with cylinders attached, nearer together and reseating one of each pair of wheels on a prepared inner wheel seat. Fig. P40 shows *Robin Hood* after conversion; patches welded into the spaces at the bottom of the tanks formerly occupied by the wheels in the "broad gauge position" can be clearly seen. At a later date the weatherboard was transferred to the bunker and an "Avonside" type front cab fitted. The latter could have come off either *Maid Marian* (fig. C174) or *Friar Tuck*, both of which later acquired overall cabs. *Friar Tuck* also received longer tanks, in 1891.

C85 — Column 2, line 59: *Raven* was reported to have been last seen in a Cinderford metal merchant's yard.

C86 — Column 1, line 27, read in conjunction with "Names", Column 2, page D43, Part 3: The late C. J. Alcock, a most reliable observer, confirms seeing 2-4-0 No. 76 with the name *Wye* when newly out of shops (to stock 11/1895) and also seeing No. 1359 with the painted name *Wye* on 19th April 1897. He also found it so described in a list dated 28/11/02 but in a list dated 10/2/09, the name had been crossed out. Fig. C176, between pages C88/9, does not seem to show a name which must either have been painted out or had worn off by the time the picture was taken.

C86 — Column 1, lines 39/40: Delete footnotes * and ‡. Vulcan Foundry supplied their Nos. 1140-59 to the Lancashire & Yorkshire Railway and their Nos. 1310/11 to the Dublin, Wicklow & Wexford Railway. Maker's Nos. 1163 and 1309 are therefore correct for *Forester* and *Gaveller* respectively.

C86 — Add after paragraph five in column 2: The map in Part 3 was a general guide to the extent of the absorbed companies' systems but the ramifications of the Forest of Dean's railways justify the inclusion of Fig. P42. With the working-out of coal reserves, traffic started to decline after World War 1 and passenger trains to all destinations north of Lydney Town disappeared after July 1929, except for excursions from Parkend; there was one from Parkend to Porthcawl as late as July 1953! The better-patronised cross-Severn service between Lydney Town and Berkeley Road came to an untimely end one foggy night in October 1960, when one of the Severn Bridge piers was demolished by an oil tanker. The track east of the river is still in use for freight traffic but the only other part of the system to survive is the four mile Lydney Junction to Parkend section, preserved by the Dean Forest Railway, see page P8.

C87 — Column 2, line 29: *Llandinam* was supplied new to David Davies at Llandinam, near Llanidloes, and could well have been used on the construction of the later-abandoned Llanidloes-Llangurig section of the Manchester & Milford Railway. It is not mentioned in M.&M. minutes. Manning Wardle records show that it was in the possession of the Carmarthenshire Iron & Coal Co. at a later date. This Company had a colliery known as Glanlash near Cross Hands in what now is the County of Dyfed but it is not known if the locomotive worked there.

C89 — Column 1, line 25: The locomotive *Bobs* at Malton Colliery was Hudswell Clarke's No. 176, not 175. The maker's records show their No. 175 (G.W. No. 1379) as with D. Davis & Son in 1917 but this entry should be treated with reserve as No. 174 was sent new to D. Davis & Sons, Ferndale Colliery, Pontypridd and the reference to No. 175 might have been a wrong entry. It seems unlikely that there could have been two locomotives in Davis' possession with consecutive maker's numbers, one new and one secondhand.

C89 — Column 2, line 16: The nameplates *Ringing Rock* were transferred to Manning Wardle's No. 890 of 1883, presumably when that locomotive was acquired by the Hundred of Manhood & Selsey Tramway in 1922 (and which changed its name to West Sussex Railway in 1924). Thus the name of ex-G.W. No. 1380 may not have been changed to *Hesperus* until 1922. It had lost its second name by May 1937. (No. 1380 was bought from Bute Works Supply Co. in 1914).

C90 — Column 2, line 22: No. 101 was Slaughter, Gruning & Co.'s No. 438 of 1861, with 15½" x 22" cylinders and 5' 3" coupled wheels. This locomotive was next recorded at G.K.N., Dowlais, where pieces were found in the 1900's.

C93 — No. 1304 *Plynlimmon* had lost its name by 25/4/10 but No. 1306 *Cader Idris* still retained its nameplates in 1913.

C93 — Column 2, line 5: C. D. Phillips, the Newport dealer, advertised a 2-4-0 passenger locomotive by L.N.W.R. on 2nd April 1900. There could have been other ex-L.N.W.R. 2-4-0's on the market but it is at least a coincidence that he lists such a locomotive in the month that the M.&M. disposed of theirs; some dimensions vary from those quoted on page C93 but this is not uncommon in sales advertisements. The price had been dropped in his January 1902 list and there is no entry in his August 1902 list.

C94 — Column 1, lines 14-16: Nos. 2301/51, 2532 were transferred on loan to the M.&M. with effect from 30th June 1905 but remained in G.W. stock. They were renumbered 8-10 to avoid objections by the Cambrian Railways to G.W.R. locomotives entering Aberystwyth station without formal running powers (which were vested in the M.&M.). From Fig. C196 it seems as if the illusion was completed by painting out the lettering on the side of the tender.

C95 — Foot of column 1: Peckett's No. 444 was built in 1885.

C95 — Column 2, line 14. *Looe* lost its nameplates in 1910/11 and its vacuum brake in 1914.

C96 — Column 2, line 3: *Pioneer* was sold to the Bute Works Supply Co. in June 1915 and it may be more than coincidence that the Horseley Iron Company's minutes recorded the arrival of a locomotive so named on 15th July 1915. This is known to have been a Manning Wardle but as the makers do not record any requests for spares for their No. 1368, the possibility cannot be confirmed in this way. It has long since disappeared and no further details have been traced.

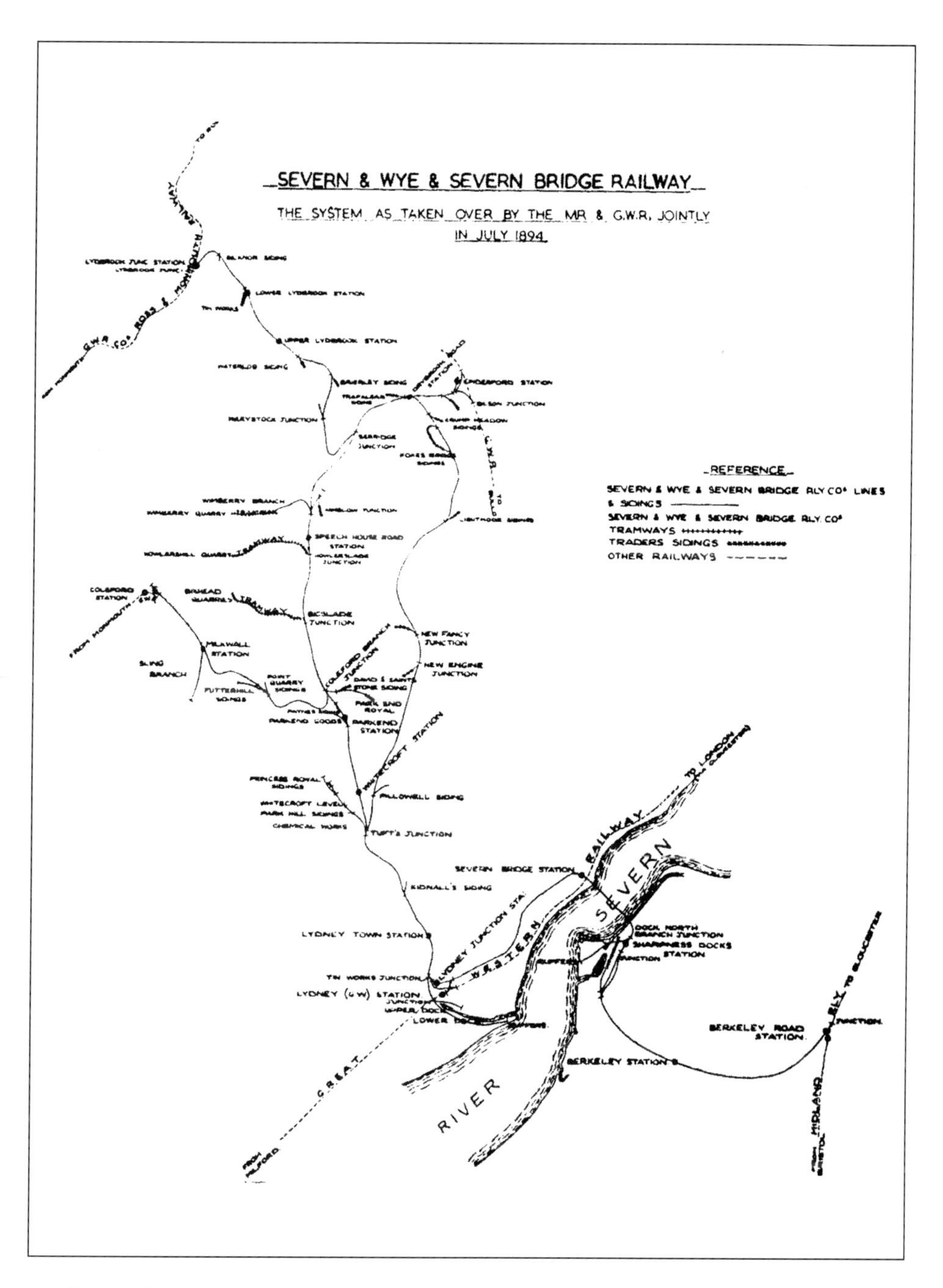

Reproduced from the Journal of the Stephenson Locomotive Society.

MISCELLANY — 1

Photograph] [W. Potter

No. 1413 at Gloucester in June 1952 with smokebox number plate but retaining letters "G.W.R." P43

Photograph] [A. G. Wells

No. 15106 at Old Oak Common in June 1954. Figure "6" picked out on number plate and painted on back of cab. P44

Photograph] [From Stephenson Locomotive Society Collection

No. 2126 fitted with fire-fighting apparatus at some time between 1906 and 1910. P45

THE RAILWAY CORRESPONDENCE and TRAVEL SOCIETY

With the Compliments of

the

Hon. Asst. Publications Officer

Ken Davies joins us in expressing thanks for your contribution to Part 14 of THE LOCOMOTIVES OF THE GREAT WESTERN RAILWAY and the enclosed copy is an acknowledgement of this.

6 Cherry Lane,
Hampton Magna,
Warwick. CV35 8SL
Tel: (0926) 402651

GEOFF BAYLIS
HON. ASST.
PUBS. OFFICER

C96 — Column 2, line 3: *Mermaid* is shown in Manning Wardle records as later with Mersey Coal Elevators Ltd., Birkenhead (entry undated but probably by about 1920) and shown in Industrial Railway Society's Pocket Book G57 as scrapped or sold after January 1929 with a query" . . to Point of Ayr Collieries Ltd., Flintshire . .?" but I.R.S. do not repeat this in their Point of Ayr list.

C96 — Column 2, lines 18 and 23, re *Nipper*: The G.W.R. register of Private Owner locomotives shows that No. 112 started work at Newport in 6/14 although the actual plate was not fitted until 1920; and *Nipper* was later with A.R. Adams, Locomotive Dealer of Newport, Mon. (now Gwent) and was seen, partly dismantled, in his yard on several occasions between January 1944 and July 1948.

C96 — Column 2, line 35: Manning Wardle records for the 1900-05 period show their No. 1040 as being with C. H. Walker & Co. Ltd., Sudbrook Shipbuilding Yard, near Chepstow. This firm took over T. A. Walker's South American interests after his death and seems to have arranged the return of several of his locomotives to this country. Delete the paragraph contained in lines 43-49.

C98 — Delete footnote (c). The maker's number 1309 on page C97 is correct.

PART 4 — SIX-WHEELED TENDER ENGINES

Page

D14 — Column 2, line 14: A contributor who travelled frequently in the Oxford area between 1900 and 1905 cannot recall Nos. 1119/32 running without names.

D16 — Column 2, line 45: *Cobham* nameplates are shown on a photograph of No. 162 with the boiler it carried from 1885 to 1896. A contemporary observer reported travelling behind No. 162, so named, in 1895 and said that the nameplates were put on in 1893 to honour a former Deputy Chairman of the G.W.R.

D17 — Column 2, line 18: The heating surface of No. 9's original boiler was 1128.0 + 146.0 sq. ft., total 1274.0 sq. ft.

D23 — Footnote*: No. 377 was stopped for repairs at Wolverhampton on 1/7/04 and was taken out of stock in 4.w.e. 20/8/04 so the alternative date of 8/03 clearly does not apply.

D35 — Column 1, line 42: Removal of No. 3201's nameplate was reported in *Locomotives & Railways* for August 1902.

D54/5 — Table — SUMMARY OF LOCOMOTIVES: In view of the complicated renumbering, this table should be read in conjunction with the smaller table and its footnote on page D53. In the table on page D55, No. 358 was built in 4/1866, not 4/1886. At foot of column 2, No. 35's mileage was 265,221 not 262,221 miles.

D75 — Column 1, lines 40-3: Nos. 2408 & 2573 were returned from loan 1/18, not 1917 and No. 2429, as well as No. 2409, was loaned to the M.&S.W.J.R. (see Page M38, Part 12).

D75 — Column 2, line 11: The movements of these locomotives are more complicated than was realised. They first were transported to Salonika (Greece) for use in Macedonia by the British Salonika Force. They then moved to Turkey over the Jonction-Salonique-Constantinople Railway for use by the British Army in Constantinople (now known as Istanbul), on the Chemin de Fer Ottoman d'Anatolie. Figs. N6 and N8, between pages N22 and N23 in Part 13, show locomotives of the class being ferried across the Bosphorous from Sirkedji, the terminus on the European side, to Haider Pacha in Anatolia on the Asiatic side.

D77 — Column 2, lines 41/2: Add No. 2460 to the list of smokebox number plates; Nos. 2409 and 2543 had lugs only at withdrawal.

D82 — Column 1, line 46: Nos. 2288, 3204/6 also had ex-R.O.D. tenders and some of the class which previously had them were later noted without. The R.O.D. tenders did not have water pick-up apparatus but as the 2251 Class rarely, if ever, worked express passenger trains over routes with water troughs, this was not a handicap.

D82 — Column 1, line 53: Nos. 2206/12/13 also had automatic staff-exchanging apparatus. Nos. 2211-14 were given old pattern 3,000 gallon tenders when built to facilitate fitting of this apparatus.

PARTS 4 - 11. NUMBERING OF GREAT WESTERN LOCOMOTIVES AND RAILCARS IN BRITISH RAILWAYS STOCK

The Great Western Railway had always been able to keep its identity, particularly in the 1923 Grouping, when alone among the amalgamating companies which formed the four Groups, it retained its name. This was neatly summed up in a contemporary cartoon showing a porter with G.W.R. peaked cap jumping for joy and saying "never even blew me 'at off". History repeated itself on formation of British Railways in 1948, albeit in a more subtle way. If all the locomotives were to be numbered in one list, inevitably those of three out of the four groups would have to be given new numbers. Great Western steam locomotives kept theirs because, it was said at the time, of the expense of removing the brass plates (the other three groups used painted numbers). The chapter on liveries shows that, initially in the case of the larger passenger classes and later of all passenger classes, the illusion that the G.W.R. lived on was to a great extent maintained.

When British Railways was formed with effect from 1st January 1948, there was the usual quest for a corporate identity but while this was being worked out, an interim renumbering system was started, wherein a letter identifying the owning Region was painted above or below the number, "W" in the case of ex-G.W. locomotives. The first example was No. 4946W, as early as 12/1/48; the practice continued at Swindon and Wolverhampton until about 16/3/48 and until early April at Caerphilly. The first examples were in yellow paint, changed to white, and below the number plates except that in the case of an ex-B.P.&G.V. locomotive, No. 2176, which (like all other absorbed locomotives) had the letters G.W.R. cast on the plates above the number, the "W" on the plate had been picked out and No.1412 had the "W" added to the buffer-beam number. There is no known official "W" record but after rejecting some obviously faulty reports there are reliable observations of 152 locomotives so treated. These included Nos. 33/52/78/91W which had been omitted from the Rhymney Railway chapter in Part 10.

Renumbering as such under the B.R. scheme only affected G.W.R. No. 2 (see later) but all B.R. steam locomotives were to receive cast iron numberplates bolted to lugs welded on the smokebox door. Although this decision was announced in mid-March 1948, it was May before Western Region first used them; moreover, they were in brass, not cast iron, no doubt to continue the tradition of the "Great Western" cab-side number plates, but this did not find favour with the Railway Executive and they were confined to "Castles" Nos. 4089/91, 5010/21/3 and (newly-built) 7009-17, to "Halls" Nos. 5954 and 6910 and to "Kings" Nos. 6001/9/25/6. Except for Nos. 7009/14-7 which were finished in standard G.W. green, they were turned out of Swindon with experimental light green, black and blue liveries respectively. Except for No. 7017 all these had smaller numbers than those which were adopted as a standard and most, if not all, eventually were painted white, with No. 4091 (Fig. P32) as an early example. The standard plates were not used on new locomotives until October, part-way through two batches. Indeed, No. 7439, the first to have the standard plate, was the last of the 7430-9 batch and was put to stock on 14/10/48, followed by No. 4166 of the 4160-9 batch, to stock 18/10/48. Nos. 6760-9 were the first complete batch to be built with front number plates.

Plates were fitted to older locomotives from September onwards, as they passed through the shops and most had been done by 1953 although odd examples did not receive plates until 1954/5. None of the locomotives withdrawn in 1948 received plates and of the 1949 withdrawals only Nos. 2942/87, 3341 and 3432 had them, while No. 1912 had been fitted with lugs only. In 1950 there were 18, plus No. 76 which ended up with lugs only, and not much progress had been made by 1951, where only 58 out of the 127 withdrawals had plates. The 4-6-0's withdrawn from 1951 onwards all had plates, reflecting their more frequent visits to shops than their humbler brethren. The situation improved, only 16 being withdrawn without plates in 1952, plus Nos. 1150 and 2181 with lugs only; twelve without plates in 1953, plus Nos. 2409, 2543 and 3160 with lugs only; six without plates in 1954 and one only (No. 2165) without in 1955. Nos. 1912 and 3160 never had smokebox number plates but Nos. 76, 1150, 2181, 2409 and 2543 may have had them and lost them. From 1956 onwards all withdrawals had been fitted with plates.

The lugs were attached to the boilers and the front number plates had to be changed each time a boiler was changed. There were several cases of locomotives running with lugs only although plates had been known to have been fitted originally and this could have resulted either from breakage of the vulnerable cast iron plates (examples of which had been noted) or even from a plate being mislaid during a boiler change. In later days there were cases of smokebox plates being removed and buffer-beam numbers restored in connection with the working of special rail tours but there were other cases where this had been done without official authority. There also were some inevitable overlaps; for example, No. 2516 was given a plate in 11/48 but retained buffer-beam numbers, while No. 6862 had a plate and a painted "W" at the same time. No. 1413 (Fig. P43) was another variant, combining a smokebox number plate with "G.W.R." on the tanks, in June 1952.

B.R. had allocated the 10000-19999 series to internal combustion locomotives and this included the only example of renumbering of a G.W. locomotive, the diesel-electric shunter No. 2 (which the G.W.R. had intended renumbering 500) and which became No. 15100. Somewhat similar diesel-electric locomotives on order as G.W. Nos. 501-7 were not completed until the B.R. scheme came into force and received numbers 15107/1-6 in the same order. These seven had G.W. style number plates but "15100" was painted on. Nos. 15101-6 went new to Old Oak Common, mainly for shunting Acton Yard where they seem to have been known for traffic purposes as Nos. 1-6, as only the last figure of each number was polished and, at least in the case of No. 15106, a large figure "6" was painted on the back of the cab (Fig. P44).

Although the two Gas Turbine locomotives had been conceived in G.W. days, no G.W. numbers seem to have been allotted. They were not delivered until B.R. days and always were Nos. 18000 and 18100.

The diesel railcars were not renumbered, either in locomotive or coaching stock, but evidently were regarded as part of the coaching fleet as their numbers were at first prefixed "W" to indicate Western Region ownership but later on some of them, including Nos. W4W and W5W, rejoiced in this double-barrelled treatment, the second "W" showing the Region to which the cars were allocated. The auto trailers described in Part 11 were treated in the same way.

PART 5 — SIX-COUPLED TANK ENGINES

Page

E43 — Note (3): No. 1741 was fitted with its new frames at Gwaun-cae-Gurwen. It was still there in 1944 but had reached Crynant by June 1947.

E46 — Sales: The Bute Works Supply Co.'s catalogues for September and October 1911 included under reference "850" a saddle tank with dimensions similar to the "850" Class. More information is to hand regarding Nos. 1923/66, 2020 (see also page M76, Part 12): No. 1923 was reported at the Ocean Coal Co.'s Treorchy Colliery early in 1944 and had been transferred from Blaengarw to Newlands Colliery, Pyle, by July 1950. In 1951 it moved again, from Newlands Colliery to N.C.B. Maesteg. No. 1966 was still at Risca in June 1949, was on loan to Blaendare Colliery in 1954/5 and was seen near Hafodyrynys in July 1958 with replacement "1966" number plates which could have been made by A. R. Adams of Newport as it also is reported as carrying a plate recording rebuilding by this firm. No. 2020 was in use at Trimsaran in July 1948 and was scrapped or sold about 1957.

E46 — Column 2, line 12: No. 2004 also had a warning bell for use at Birkenhead Docks.

E47 — Column 1, line 24: No. 1912 left Wolverhampton in 2/49 with lugs only and never carried a smokebox number plate.

E49 — No. 1991, condemned 1/53 was retained for stationary use. It was noted at Briton Ferry on 13/10/53 and had reached Swindon by 24/4/55.

E50 — Line 27 (footnote): No. 1919, withdrawn in October 1938 was fitted with a Belpaire boiler and an overall cab in preparation for reinstatement in May 1940.

E52 — Column 2, line 32: No. 2034 also had a spark arrester. Nos. 2101/44 had been so fitted for work at the British Sugar Co.'s factory, Kidderminster.

E52 — Column 2, line 39: Nos. 2082 and 2146 had steps welded to the bunker on the fireman's side as described for No. 2092 on page M76 and Nos. 1709 and 1888 were not unique, as claimed on page E64, in column 2, line 56.

E52 — Column 2, line 51: No. 2052 also had a warning bell when at Birkenhead.

E52 — Foot of Column 2, Later Details: No. 2126 (fig. P45) was fitted with fire-fighting equipment comprising a firehose cabinet at the back of the bunker and a steam-driven pump between the leading and driving wheels. It is not clear if the apparatus was only able to draw upon the limited storage capacity of the saddle tanks or whether the hosepipes below the footplate could be connected to hydrants situate around the shed yard. The photograph was taken at Old Oak Common shed, where No. 2126 was stationed from 1906-1910 and it seems likely that this was the period in which the apparatus was available for use, as No. 2757, similarly fitted (see later) was fitted with pannier tanks in July 1911, when the apparatus may have been removed, although there are no written records of the fitting to either No. 2126 or 2757.

The foot of Column 2 also refers to the modification and renumbering of ten of the class into the 2181-90 series. No new Diagram was issued, presumably the modification of the brake gear was not regarded as significant enough and the locomotives remained on Diagram B52 in the 0-6-0T series.

E53 — Column 2, lines 3-5: Delete No. 2048 from, and add Nos. 2086 and 2190 to, the list of "2021" Class fitted with smokebox number plates.

E53/4 - Nos. 2038/42, both condemned 4/53, were seen intact at Swindon on 30/7/56 and presumably had been used as stationary boilers, although both officially recorded as cut up in 1953!

E54 — Table: Under *Pannier Tanks* alter the date 12/19 for No. 2061 to 9/32. 2061 was photographed as a saddle tank in 1932 and would have received pannier tanks along with the B4 type boiler. In January 1952 it was stated that No. 2066 (condemned 9/51) had been retained for use by the Engineer's Department for track removal in the Cinderford area, which presumably was completed by the time it was sent from Swindon to Southall Shed for use as a stationary boiler in 7/52. Returning to Swindon 3/54, it was cut up in 5/54.

E56 — Line 4: No. 2128 probably did little, if any, work after the arrival of ex-G.W. No. 789 (page K183) in 1930.

E57 — Alter lines 52-4 to read "Nos. 1814/23 were turned out from Swindon in November 1905 with B4 type boilers" (No. 1823 was *not* done at Newport).

E60 — No. 1833 was weighed on completion on 9/11/82, weight recorded as 42T 10C.

E61 — Column 2, line 6: It seems possible that the Bute Works Supply Co. of Cardiff acted as G.W.R. agents in seeking to dispose of some members of this unpopular class because their catalogues for September and October 1911 described a saddle tank locomotive of this class, even down to quoting the exact heating surface (1326.74 sq. ft.) of a Group 27 boiler.

E69 — Column 2, line 40, Later Details: No. 2757 (fig. P46) was fitted with fire-fighting equipment similar to that described for No. 2126 above. No. 2757 was stationed at Old Oak Common shed from December 1906 until April 1911.

E75 — Column 2, line 38: Boilers of Appendix Group 85(e) were standard for the class as built but there was a late fitting of four boilers of Group 85(f), Code BC, to Nos. 5625 (3/60), 5668 (11/59) and 6671 (10/59 & 2/63). Like the 2-6-2T Nos. 4160-79 (Page J30, Column 2), built with this type of boiler they had large pressings covering the oil pipes on the right-hand side where the barrel and smokebox joined.

E76 — Column 1, line 25: No. 6692 did *not* have a boiler from a "rebuilt absorbed

MISCELLANY — 2

Photograph] [V. R. Webster's Collection

No. 2757, fitted with fire-fighting apparatus, at Old Oak Common Shed at some time beween 1906 and 1910. P46

Photograph] [LCGB Ken Nunn Collection

No. 6166 leaving Marylebone on the 11.20 a.m. to Princes Risborough in March 1950. P47

Photograph] [P. J. T. Reed's Collection

45XX 2-6-2T No. 2185 (4524) at Brent. Note non-extended smokebox and absence of front struts, like No. 2173 (Fig. J75). P48

Photograph] [Reproduced by courtesy of the National Railway Museum, York

No. 3297 *Earl Cawdor* **near St. Mary Cray, Kent, on a non-stop run with Queen Victoria's Royal Train from Windsor to Folkestone on 11th March 1899.**

P49

0-6-2T". It retained its original boiler (No. 5812) until 9/51. Repairs to tubeplate and tubes were made at Caerphilly in 1945 and the cast iron taper chimney presumably was fitted at that time, as it was reported in the *Railway Observer* in November 1946. The photograph of No. 6692 with a similar chimney which forms the subject of fig. E120 (facing page E65) would have covered the period of a different boiler, No. 4155, carried from 11/51 until 6/55. There is now no means of telling whether the chimney was transferred from Boiler No. 5812 to No. 4155 or whether it was another non-standard fitting but its existence is confirmed by a second photograph, taken on 14/6/53.

E78 — Column 2, line 39: In 1960, Nos. 6741/69 were noted with screw couplings at the front end only.

E80 — British Railways: Following the transfer of Nos. 1367-9 (pages E75 and M77) to Wadebridge in the summer of 1962, the Weymouth Quay line was worked by 57XX Class locomotives and the last steam-hauled Channel Islands boat train was worked by No. 4610 on 24th December 1963.

E80 — Column 1, line 51: London Transport lined maroon livery is shown on Fig. P37.

E80/1 - SUMMARY OF LOCOMOTIVES: Alter building dates of Nos. 7705-13 from 3/30 to 7705-9, 3/30 and 7710-13, 4/30; and Nos. 9782-4 from 7/36 to 9782, 6/36 and 9783/4, 7/36.

E85 — Column 1, line 22: Nos. 7400-49 had lever reversing gear whereas Nos. 6400-39 had screw reversing gear. The latter was particularly necessary in auto working when the driver was controlling the train from the drive-end trailer and the fireman had to "notch up" to adjust the cut-off.

PART 6 — FOUR-COUPLED TANK ENGINES

Page

F8 — Further research has been made into apparent discrepancies between G.W.R. and Sentinel Waggon Works' records and it appears that Nos. 12 and 13 were identical when first delivered except that No. 12 was fitted with vacuum brake and steam heating for passenger work. Both had the smaller boilers for which dimensions are tabulated on page F9.

The G.W.R. C.M.E.'s report for 1926 refers to trials on the Malmesbury branch in late September and October 1926, stating that the limited boiler capacity could not meet traffic requirements. The trials were terminated and No. 12 transferred from Swindon to Park Royal, from where it was set to work on the Brentford branch. Taken out of stock in four weeks ending 25/12/26, it was returned to Shrewsbury on 13/1/27, where the frames were altered to accept a larger boiler, 25% bigger than the original. After Works trials, "6515" was loaned to the Shropshire & Montgomeryshire Light Railway for passenger train trials and next appears to have returned to the G.W.R. for the further trials on the Malmesbury branch, on 2nd October 1927 (described on page N31, Part 13). Returning again to Shrewsbury, it did duty as a works shunter and is also reputed to have gone on loan to a China Clay works in Devon before sale in October 1934 to T.E. Gray. They sold it in turn to a consortium of six at Quainton Railway Centre in May 1972, where it arrived on 31st May and was restored to working order in August 1979.

The situation is further confused because the official Diagram mentioned on page F8 shows the larger boiler, which was not fitted until after it had completed its first stay on the G.W.R. and was not representative of No. 13. The latter was condemned 27/5/46 and sold in June, whereupon Swindon Drawing Office ruled a diagonal line across the Diagram entitled "Sentinel Locomotive No. 12" with the legend "Condemned June 1946"!

F19 — SUMMARY OF LOCOMOTIVES: Delete "3/08" against No. 1467 which never was converted to long wheelbase.

F19 — Footnote (e) and Fig. F30: Fig. P16, taken at Craven Arms in 1906 shows No. 567 in its ex-G.W.R. condition and Fig. F30 shows the only visible B.C.R. modification, the enclosed cab. The latter was taken in May 1932 and there is an earlier picture in existence, taken in 1931, showing the same cab.

F30 — Column 2, Condensing Apparatus: Official information to hand confirms that No. 459 was built without the apparatus and that Nos. 978/80/5/6 received it in 4/89, 8/90, 11/90 and 8/89 respectively.

F40 — Column 2, Allocation and Work: line 45 refers to No. 3602 moving south; Nos. 3602/3/10 were all shedded at Kidderminster between June and December 1920 and Nos. 3605/10, see line 50, were noted at Kidderminster shed in 1927.

F42 — Column 1, line 52: The tank capacity was 850 gallons and when No. 1 was weighed on completion in May 1882 the weight was recorded as 45T 18C.

F44 — Column 2, line 45: The experimental firebox fitted to No. 1490 was of steel.

F46 — Column 1, line 10: Another official record shows the boiler as fitted in June 1903 and as the locomotive was added to stock in 4.w.e. 27/6/03 the date of fitting the Lentz boiler in the summary should be altered to 6/03. No. 101 had clearly been in service for a month or so as it went back into Works from 6/8/03 until 20/10/03.

PART 7 — DEAN'S LARGER TENDER ENGINES

Page

G7 — Column 1, line 4: Add No. 4122 to the list of "Atbaras" with steel fireboxes.

G9 — Column 1, line 17: Delete "(possibly a misprint)". No. 3005 had a bogie with volute springs when weighed in September 1897 and was reported thus in the May 1899 *Locomotive Magazine.*

G12 — Footnote (o): the nameplates *John G. Griffiths* were not fitted to No. 3412 until 7/14, see footnote (h) page G27, so that they may have remained on No. 3060 until that date. One Swindon record says removed 3/14, another gives 7/14.

G14 — Column 2, line 25: No. 3254 retained straight cab sides and small tender in September 1947; No. 3265 received curved cab sides between August 1939 and 1949; No. 3273 had curved cab sides by August 1939.

G17 — Remove No. 3280 from footnote (a) and transfer to footnote (b).

G17 — Revise footnote (e) to read: The withdrawal date against No. 3265 on page G16 refers to the former No. 3365, renumbered 3265 (and later 9065). The original No. 3265, according to its History Sheet, was stopped for repairs at Swindon 11/12/29, condemned January 1930 with a mileage of 918,405 and cut up February 1930. (Also corrects line 33, column 2, page G14).

G18 — Column 1, line 31: Nos. 3200-18/20-8 had the cab side sheets set outwards but No. 3219 was the exception, retaining the original straight sides in September 1947. It was condemned in November 1948.

G20 — Column 2, line 9: Transfer of the mechanical stoker to *Ladysmith* was recorded in *The Locomotive* for November 1904.

G25 — Line 9: The History Sheet for No. 3365 says it was sent to Swindon for heavy repair on 17/12/29 with mileage 864,252, converted from "Bulldog" to "Duke", renumbered 3265 and renamed *Tre Pol and Pen*, 31/12/29. Boiler No. 3285, "113 Class" replaced No. 4138 "Standard No. 2". (This replaces the two lines at the foot of page N31. The note at the top of page N32 still applies).

G25 — An official list of intended names shows No. 3367 (old 3419) as *E. H. Llewellyn* and the magazine *Locomotives & Railways* of April 1903 reported it as such, yet the *Locomotive Magazine* of 21/3/03 reports it as a "new engine *Evan Llewellyn.*"

G27 — Footnote (b): The combined oval name and number plates were removed from No. 3353 *Plymouth* on 21/4/27 and as the locomotive was in use until nameplates *Pershore Plum* were fitted on 7/5/27, a pair of standard numberplates must have been fitted on or about 21/4/27. Although No. 3384 (then No. 3446) was taken into stock in September 1903, it did not leave Works until 12/10/03 and may never have carried the *Liverpool* nameplates.

The policy behind the removal of nameplates "to avoid confusion with train destinations" (see also pages G17, footnote (b) and G36, footnote (m)) was illogical as many of them were not even stations, while none of the "Cities", Nos. 3710-9 lost theirs, neither did No. 4070 *Neath Abbey*, which *was* a station between Neath (Riverside) and Swansea (East Dock).

G27 — Footnote (h): *Joseph Shaw* did not become a Director until 7/12/17 and as No. 3434 was stopped for shops in October 1917 and did not leave Swindon after overhaul until March 1918, (10/17) should be altered to (3/18).

G27 — Footnote (k): The *Locomotive Magazine* for 26/9/03 reported Nos. 3381-3 (old Nos. 3443-5) as having been turned out without nameplates but the 17/10/03 issue said that No. 3443 had been named *Birkenhead.*

G30 — Column 1, line 8 and G33, column 2, line 19, steel fireboxes: Official records do not give the material of 3310's firebox but a contemporary account says it was of copper. Seven, not five, of the "Atbaras" had steel fireboxes.

G31 — Allocation and Work: No. 3297 *Earl Cawdor*, the tender provided with extra water capacity for the long run, worked the Royal Train non-stop from Windsor to Folkestone Pier on 11/3/99, conveying Queen Victoria on her journey to Nice. Fig. G49 shows the train in "foreign territory" at St. Mary Cray, Kent.

G33 — Column 2, line 38: No. 4138 (old 3392) was fitted with Westinghouse brake equipment when built as the official records show it taken off 2-4-0 No. 3249 in 1900 and put on No. 3392. No. 4138 was condemned in 11/29 and the equipment reputed transferred to No. 6332, so fitted in 3/30, Part 9 page J14. The reference to this equipment on *Powerful* on page G33 was based upon the picture No. G53 in Part 7 but there is nothing in the written records and no other picture of *Powerful* so fitted. Picture G53 may be of No. 3392 temporarily renamed and renumbered *Powerful*, No. 3385, for the C.I.V. special trains on 29/10/00 (see footnote (c), page G36). It will be better to ignore the reference to this equipment on page G33 and in the caption to Fig. G53. One of the duties of the locomotives so fitted was to take over horsebox traffic from Newmarket at Acton.

G36 — Footnotes (a), (b) and (c): The three locomotives renamed had combined name and number plates and were automatically "renumbered" when the plates were changed although the official Working Notice of 28/10/00 refers to these three by their changed names and their original numbers!

G36 — Footnote (d): Figure G50 in Part 7 is an official photograph, taken in May 1900 and showing No. 3375 *Edgcumbe* in "shop grey". It would have had to go back into shops for final painting, therefore the name *Conqueror* cannot have been carried in service. *Locomotives & Railways* for May 1901 said that No. 3375 "now has its name over the driving wheels" and there is a picture of a very short arc nameplate *Edgcumbe* attached to the boiler barrel about six inches above the splashers. Another photograph dated 10/7/03 shows it as *Colonel Edgcumbe* with a traditional arc nameplate, which was reported in the *Locomotive Magazine* for July 1903.

G40 — SUMMARY OF LOCOMOTIVES: Nos. 2608/10 were withdrawn respectively in 12/05 and 3/06, not 12/06 and 8/06, being replaced by "Aberdares" in 1/06 and 4/06.

G42 — Column 2, Allocation and Work: While still in stock, No. 2625 was "allocated" to a training area in West Wales in the spring and summer of 1945, where it was painted white and used for target practice by the United States Air Force.

PART 8 — MODERN PASSENGER CLASSES

Items common to 1st and 2nd Editions

Page

H8 — Both Editions, column 2, line 38: A photograph of No. 4057 taken in September 1919 shows the brass beading intact, although it was removed at a later date.

H9 — Column 1, line 44: Nos. 4004/15/28/44 also received speedometers but there is no confirmation of a published report that one was fitted to No. 4014.

H9 — Column 1, line 63: No. 4042 had the later type of "Castle" chimney on 19/5/44.

H9 — Column 2, ALLOCATION: Doubt about a Cardiff allocation in 1913 has been resolved by official confirmation that Nos. 4001/8/14 were all there between 1909 and 1913. Fishguard had a brief life as an Ocean Liner Terminal between 1909 and 1914, when special mail and passenger trains were run. With one stop, at Cardiff, the Fishguard-Paddington journey was run in about $4^{1/2}$ hours.

(In the 1992 timetable, with the advantage of High Speed diesel trains (HST's), the journey still takes four hours, albeit with three stops!). No. 4019 went to Fishguard for these duties in March 1910 and at the climax of the workings in the late summer of 1914, there were six "Stars" shedded there, with three remaining for the Irish traffic after the Ocean Liner specials ceased to run.

H10 — SUMMARY OF LOCOMOTIVES: No. 4019 did not receive elbow type outside steam pipes and the date 5/48 should be deleted.

H11 — Both editions, footnote (a): One Swindon record gives the date of naming No. 40 as June 1906 yet it was not reported in the railway press (in three magazines) until November and December, the first being the *Locomotive Magazine* of 15/11/06 and included the G.W.R.'s own Magazine for 12/06. No. 40 was stopped at Swindon from 10/10 to 1/11/06, so that October 1906 seems a more likely date of naming than the 9/06 given in footnote (a) or the official 6/06.

H15 — Column 1, line 37: Nos. 7008-17 definitely came out without speedometers and Nos. 7018-27 probably lacked them. They started to be fitted in 1950 and all had them by 1954.

H19 — Footnote (a): The South Wales Borderers have a name and numberplate from No. 4037 in their Museum in Brecon but no regimental plaque. No. 4037 was photographed with the plaque in place on the left-hand side on 6/7/62 while a right-hand side photograph taken on 9/7/62 shows a blank splasher with holes where the fixing bolts would have been. The left-hand plaque must also have disappeared by the time 4037 was withdrawn on 17/9/62.

H20 — Both editions, "6000" Class, paragraph 2: The class was restricted, as stated, to the main lines from Paddington to Plymouth and Wolverhampton and, while some trials were carried out in 1938 up to twenty miles west of Plymouth, the first moves to extend the route availability came in 1948 when "Kings" were allowed between Bristol, Shrewsbury and Wolverhampton and three were stationed at Bristol (Bath Road). A "King" was on trial between Shrewsbury and Chester in December 1949 and the class was allowed from the end of January 1950 but there are no records of any regular workings. Following a multiple failure of new diesel locomotives one day in the changeover period of the late 1950's, No. 6026 *King John* was used, pragmatically rather than officially, to work an express train over Saltash Viaduct, from Plymouth to Truro. It returned the same day, running light, under cover of darkness. The only other significant relaxation in steam days allowed "Kings" to work Paddington to South Wales expresses as far as Cardiff but more recently, upgrading for increased axle loads of freight wagons has made it possible to run "Kings" on special trains as far west as Carmarthen, the subject of our frontispiece, fig. P1.

H20 — First Edition, Column 2, after line 19: Second Edition, Page H21, Column 1, after line 11: With the intention of distributing the increased axle loading evenly, Nos. 6000-19 were built with equalising beams between the coupled wheel spring hangers but these did not live up to expectations. Nos. 6020-9 were built without them and they were removed from the first twenty. By contrast, the springing of the bogies was made more sophisticated by fitting of coil springs to supplement the leaf springs following the partial derailment of the bogie of No. 6003 at Midgham on 10th August 1927. This applied to Nos. 6000-5; Nos. 6006-29 were turned out with this feature.

H23 — First Edition, Column 2, line 51, Second Edition H24, column 2, line 10: delete "and 3,500 gallon tenders" and substitute: "It is uncertain what tenders were attached at first but photographic evidence suggests that Nos. 173 and 176-8 had 4,000 gallon tenders and Nos. 174/5, 3,000 gallon tenders. No. 175 had acquired a 4,000 gallon tender by 8/05."

H25 — First Edition, column 1, line 36, Second Edition, column 2, line 23 should have mentioned war-time removal of brass beading as for the "Star" Class (described on page H8 in both Editions). As with that class there was the inevitable exception in that, like No. 4057 mentioned above, another photograph taken in September 1919 shows No. 2973 with brass beading intact but it also was removed at a later date.

H25 — First Edition, column 1, line 58; Second Edition, column 2, line 11: The distance pieces between nameplates and splashers were missing from No. 2931 on photographs dated 1922/3 (Fig. P69) and 8/31 but had been fitted by 6/35.

H27 — First Edition and H28 Second Edition, SUMMARY OF LOCOMOTIVES: under Boiler Type Changes, for No. 2927 "D2 10/10" should read "D2 1/09". (The editor apologises for showing 2929 instead of 2927 on page M95; the entries against 2929 in both editions still apply). Also on pages H27/8 First Edition and H28/9 Second Edition, for No. 2932 add "D2 12/15, D4 11/17"; for No. 2976 "D4 6/12" should read "D4 9/13"; for No. 2987 "D2 8/09" should read "D2 5/10"; for No. 2989 "D4 4/12" should read "D4 6/13"; for No. 2998, "D2 1/13" should read "D2 6/13". No. 2976 was superheated in 3/12, not 3/11 and No. 2979 in 8/14, not 6/12.

H28 — First Edition and H29 Second Edition, footnote (h): The date of renaming No. 2975 was quoted in error from an internal Swindon memo issued after the event. No. 2975 entered Swindon Works for a light repair as *Sir Ernest Palmer* during week ending 16th September 1933 and left on 25th September as *Lord Palmer.*

H30 — First Edition column 1, line 14, Second Edition column 2, line 15: No. 5986, to stock 21/11/39, was the first locomotive to have no cabside windows.

H30 — First Edition, column 1, line 38, Second Edition, column 2, line 38: "Hall" Class Nos. 5931-40 had the device for automatic clipping-up of the A.T.C. shoe for working over electrified lines in the London area. These special shoes replaced the standard pattern between June 1934 and February 1935.

H30 — First Edition, column 2 between lines 24 & 25; Second Edition, page H31, column 1 between lines 35 & 36, add a new paragraph: Nos. 6959-65 had modified lubricating gear (the usual five-feed lubricator with a new design of control valve) to cope with more stringent working conditions which might obtain in the cylinders due to the use of the larger three-row superheater. Nos. 6966-70 with standard two-row superheaters also had the same modified lubricating gear for comparison. Also, Nos. 6959-61/5/6/70 were fitted with auto-feed drifting gear which ensured that a mix of steam and oil was fed to the valves and cylinders during drifting periods, even if the regulator was closed completely. This was a small single-feed mechanical oil pump under the left-hand footplate and was operated by the left-hand rocking shaft.

No. 6981 came out twelve days after No. 6982 and it was said at the time that this was due to a boiler modification but the records show that Nos. 6971-90 all had Standard No. 1 Group 97(h) AK boilers numbered consecutively 9223-42 and there seems to be no reason for the statement.

H34 — First Edition, column 1, line 46; Second Edition page H35, column 1, line 23: Swindon records show 68XX as replacing 43/83XX in the order in which the latter were condemned but this list (see below) is purely nominal. No. 6800 utilised parts from No. 4313 but the components of subsequent 2-6-0's, which would have been sent to the Machine and Wheel Shops for refurbishment before re-use, were evidently not returned in the same order. The motion of Nos. 6802/10 was stamped 8322/57, both of which survived to be renumbered 5322/57, of which No. 5322 is preserved at Didcot. The motion on Nos. 6807/9/39/41/2 was stamped 4383/56/66/08, 8342 in that order but this could be pure coincidence. The list on page N33 ("J14 column 2, line 40") contains three errors and should be ignored, being replaced by:

Nos.	6800-9	nominally	replaced	Nos.	4313/09/64/31/2/97/87/3/50/6
"	6810-9	"	"	"	4395/44/10/51/84/41/5/98/01/40
"	6820-9	"	"	"	4329/72/34/23/57/76/35/46/11/24
"	6830-9	"	"	"	4347/74/8/27/70/33/91/38/48/66
"	6840-9	"	"	"	4315/08, 8342/04, 4359, 8383/54/ 89, 4396/06
"	6850-9	"	"	"	4314/7, 8352, 4300/5/52/93/30/ 73/62
	6860-9	"	"	"	4336/80/28/94/04/71/92/67/39/88
:	6870-9	"	"	"	8308, 4363/8/1/90/85/54. 8329/ 66/87

H36 — First Edition, column 1, line 9; Second Edition, page H36, column 1, line 53: The remarks above about the relationship of 68XX to 43/83XX equally apply to the "Manors" and the list on page N33 again should be ignored and replaced by:

Nos. 7800-9 nominally replaced Nos. 8363, 4322/1/16/43, 8301, 4389/82/69/79
" 7810-9 " " " 4307/12/25/19/42/60/99/02/49/55

H36 — First Edition, H37, Second Edition, BRITISH RAILWAYS: In 1953 the Swindon authorities tried to persuade the British Railways Board to sanction another ten "Manors", Nos. 7830-9 for use on the Cambrian lines but were allocated B.R. 75XXX Class 5 4-6-0's.

H39 — First Edition, H40 Second Edition and page M62, part 12, BOILER APPENDIX: Boilers of Group 97(g), Standard No. 1 coded AL and AN had larger pressings covering the oil pipes on the right hand side where the boiler barrel and smokebox joined (Fig. H59, both Editions). These boilers were fitted at various times from 1943 onwards to "Saint", "Star", "Hall" and "Grange" classes. There were a number of variations in the size and shape of these pressings through the years.

Item relating to Second Edition only

H24 — Column 2, line 2: "as" to read "are".

Cabside view of "Hall" Class 4-6-0 No. 5958 in February 1948 showing, from top to bottom:

Power Group letter "D" and Route Colour disc (Red), see Part 1, page 59; "X" to show that normal loading could be exceeded and "W", the temporary suffix painted on in early 1948 to denote Western Region locomotives, see Part 8 (First Edition) page H31 and (Second Edition) page H32.

Photograph] [R. H. G. Simpson

PART 9 — STANDARD TWO-CYLINDER CLASSES

Page

J13 — Column 1, line 51: The lengthened frames also made a longer cab possible and the changes are clearly shown on Diagrams G & H and although the angles of the photographs are oblique, can be seen by comparing Fig. J8 with Figs. J9/11-13. An equally noticeable alteration to the cab outline from Nos. 5390 onwards (see page M109, "J13 — Column 2, line 40") did not involve a Diagram change, probably because such a feature did not affect vital dimensions. The difference is shown on Figs. J9 and J10.

J12 — Table of building dates, 43XX Class: Although the parts for Nos. 7305-19 were supplied by R.S. & Co., they did not allocate works numbers 3837-51, which were used for locomotives sent abroad.

J14 — Column 1, line 23: The first twenty reversions from 83XX to 53XX were authorised to bring them into the "Blue" group for working over Southern Railway metals under wartime arrangements.

J14 — Column 1, line 48: The list of locomotives with automatic tablet changers should read 4339, 6305/23/63/4/72/83/98, 7314.

J14 — Column 2, line 39 and table at foot of Column 2, page J18: When Nos. 9301/17 were altered to Nos. 7323/39, number plates were not available and the new numbers were painted on the cab sides in "Great Western" style figures, and painted on the smokebox doors. These were soon replaced with standard patterns and the other eighteen conversions came out with 73XX smokebox number plates.

J15 — Column 1, line 59: The numbers were painted on the tenders in large figures along with the letters R O D but the numberplates were retained according to photographs of Nos. 5319 (see Fig. N7, Part 13), 5322 and 5328 taken in France on active service. They arrived painted green and were repainted black, no doubt for camouflage reasons, at Audruicq, Pas-de-Calais.

J15 — Column 2, line 11: No. 4353 did not receive a smokebox number plate.

J20 — Column 2, line 39: Nos. 2884 upwards all had the small cast iron chimney when built but No. 3832 was seen with a large parallel chimney in April 1955.

J21 — Column 2, line 22: The claim that No. 2846 rather than No. 2804 was involved in the Glenfarg incline trials is based upon a North British Railway Board Minute of 6th May 1921, four months after the event. N.B.R. 0-6-0 No. 46 worked on comparative trials and perhaps the minute-taker confused Nos. 46, 2804 and 2846. There is no doubt that No. 2804 was used as it is so described in the official report by the G.W.R.'s Chief Locomotive Inspector, C. T. Read, who conducted the trials. Moreover, No. 2846 was in Swindon Works at the vital time and did not leave there until 11th January.

J23 — Table: No. 2876 was photographed ex Works with outside steam pipes at Swindon in February 1956 after major repairs: Delete "6/59‡" and replace by "2/56".

J25 — Column 1, line 30: The mounting of the snifting valves outside the steam chests applied only to Nos. 4701-3 (not to 4704) but they were fitted to No. 4700 when rebuilt with the Standard No. 7 boiler.

J26 — Column 2, line 3: No. 2230 was weighed on completion on 17/10/06, at 76T 16C.

J29 — A feature which applied to most of the 2-6-2T, 2-8-0T and 2-8-2T described in Part Nine but not previously mentioned (and to the 56XX and 66XX described in Part Five) was the provision of a pair of handrail grips on the bar which bridged the boiler at the front end of the water tanks. These enabled the fireman to steady himself on the sloping tank top when the locomotive was taking water. The first known fittings were to No. 3100 (to stock 12/38) and No. 8104 (to stock 1/39) and the available evidence suggests that all new and reconstructed locomotives after this date were so fitted, comprising Nos. 3101-4, 4130-79, 5255-64, 7250-3 and 8105-9. At this stage it is not possible to say whether fitting to earlier locomotives with sloping tanks was universal but there is photographic and visual evidence of the feature on members of the following number series: 3150-90, 4100-29, 4200-99 (except conversions to 72XX), 4555-99, 5100-99, 5200-54, 5500-74, 5600-99, 6100-69, 6600-99, 7200-49 and

8100-3. The grips were not fitted to Nos. 4400-10 which had straight-topped tanks, but *were* to some of Nos. 4500-74 which also had straight tanks.

J30 — Column 2, J31, Column 1: The boilers with new superheaters fitted to Nos. 4160-79 had a larger pressing over the oil pipes entering the smokebox. These are shown in the picture of No. 4171 (Fig. P73, included on the same page as No. 4257 which has the same type of pressing on a different boiler). The boilers fitted to Nos. 5625/68 and 6671 (Page P54) also had this pressing.

J31 — In keeping with the Great Western penchant for single exceptions to standard arrangements, No. 5166 left Wolverhampton Factory on 14/9/46 fitted with a tapered cast iron chimney.

J33 — Allocation and Work: From 26th September 1949, Nos. 6129/66 were transferred to Neasden Shed (Eastern Region) and worked Aylesbury-Princes Risborough-Marylebone trains (fig. P47). They were out-stationed at Aylesbury Shed and continued on these duties for almost two years, being transferred back to Western Region in July 1951.

J37 — Column 1, line 3: No. 3160 never did receive a smokebox number plate.

J47 — Column 1, line 5: Photographs show that Nos. 2161/3/4 had the original small bunker, no pictures of Nos. 2165-72 as built are available, but No. 2173 (Fig. J75, Part 9) had an extended bunker when new, while No. 2176 (Fig. J78 as 4515) had a fully-sloping extended bunker. There is no evidence about Nos. 2174/5/7/8 but a picture of No. 2179 shows an extended bunker. The boiler has top-feed so the bunker could have been extended in the meantime.

J47 — Column 1, lines 15-18: No. 2185 (fig. P48) had neither an extended smokebox nor front-end struts and these features are unlikely to have applied to the other nine of Lot 174.

PART 10 — ABSORBED ENGINES 1922 - 1947

General Notes

The total number of locomotives given under the title of each Company, e.g. 39 on page K8 and 148 on page K23 represents those which had been allotted numbers in the G.W. list as described on pages 15-17 in Part One, as amended by Part Twelve, pages M64/5. Except for the T.V.R., these totals agree with Table 1 facing page 17 in Part One. In four cases, the totals in the Part One table were reduced by withdrawals prior to the date when the various companies' locomotives were added to G.W. stock. The details are:

Cambrian Railways: Total 99 on page K53 to read 78; 76 of these were added to G.W. stock in four weeks ending (4.w.e.) 10/9/22 and two in 4.w.e. 8/10/22 (Nos. 57 and 59). Those not taken into stock were the 21 shown withdrawn between 5/22 and 8/22 on pages K84/5.

Cardiff Railway: Total 36 on page K87 to read 34; these were added to stock in 4.w.e 10/9/22, Nos. 24 and 36 having been withdrawn 5/22, see page K96.

Neath & Brecon Railway: Total 15 on page K235 to read 14 which were added to stock in 4.w.e. 8/10/22, No. 15 having been withdrawn 8/22, see page K242.

Taff Vale Railway: Total 274 on page K134 to read 271. (A total of 275 had been given in Table 1, Part 1, by inclusion of T.V. No. 333 — see foot of column 2, page K155). The 271 locomotives were added to stock in 4.w.e. 8/10/22, Nos. 281/8/97 having been withdrawn in 4, 5 and 6/22 respectively, see pages K194/5.

Information on second-hand locomotives prior to their acquisition by the companies concerned or after disposal by the G.W.R. or its constituent companies is not usually available from official sources except where accurate copies of locomotive builders' lists exist. Most of it comes from the records of Societies specialising in Industrial Locomotive history based upon personal observations and research by their members and help from these sources is hereby gratefully acknowledged; but different recorders may take a different view of a locomotive's condition and dates given by workshops may not agree with a company's books and accounts. The following examples will show how precision can be difficult: No. 244, shown on page K32 as scrapped in 6/52 was lying disused in the open in Walkden Yard on 26/5/51, shown in the workshop records as "Scrapped 15/10/51", had been reduced to a chassis only by 2/52 and disappeared entirely by 26/8/52; and No. 700, Dowlais Steelworks No. 22, was reported derelict there in 8/48, converted to a transporter wagon by 7/49 and obviously finally cut up in this form in 8/50, as recorded on page K27.

More disturbing is the case of No. 579 (page K175) which became *Kitchener* and later *Wellington* on the Longmoor Military Railway. This account and confirmation of the scrapping date of 1944 (Part 12, page M118) came from two apparently reliable sources. Then two separate publications quoted date of scrapping as "in or by 1940" and this version is supported by eye-witness accounts of two people who actually were stationed at Longmoor at the vital times. There was no trace, intact or otherwise, in late 1940 or in 8/41, twice in 1942 or in 1943 or 1944.

Another valuable source of information has been the sale particulars published by locomotive dealers, notably C. D. Phillips of Newport (Mon., now Gwent) and the Bute Works Supply Co. of Cardiff (who both also are mentioned in places in Part 3). Not only did these two firms buy and sell locomotives which they actually owned but acted as agents both for potential buyers and for railway companies wishing to sell surplus stock. There is evidence that Mr. Phillips acted in this way for the Alexandra Docks company and that the Bute Works Supply Company acted not only for the B.&M. and M.&S.W.J., but even for the G.W.R. itself.

Research into fitting of smokebox-door numberplates reveals that the following did *not* receive plates: Nos. 270 (page K32), 864/73 (K73), 433 (K209), 2165/7/92/5/7 (K217). Those not previously listed which did receive them were Nos. 271 (page K32), 7, 8 & 9 (K78), 203-5/7-11/15-8/36/79/82/4/5/90/2/3/5/9 (K217), while in the case of Rhymney Railway locomotives, Nos. 33, 52, 78 and 91 received the temporary "W" suffix and Nos. 35-44/6/55/6/8/9/63/5-70/2/3/5/7-83/91-4/6 were fitted with smokebox door plates but No. 76 had lugs only when withdrawn in 11/50 (pages K119/22-4/6/8). The other ex-Rhymney locomotives described on pages K119-128 did not receive smokebox door plates.

Part 10 — Detailed References

Page

K10 — Column 1, line 34: Delete "later in the year" and replace by "in December 1900". Mr. Phillips advertised Nos. 7 and 22 on 1/8/00, apparently as an agent for the railway company but the A.D.R. themselves seem to have been the source of an advertisement in the GLASGOW HERALD for 27/10/00, when these two, together with No. 11 (see page K11) were briefly described. Mr. Phillips advertised No. 22 in February and March 1901 and it also appeared in a Bristol auctioneer's catalogue in April 1901. Phillips advertised No. 7 from February to September 1901. There is no further trace of any of the three.

K11 — Column 2, disposal of No. 10: Mr. Phillips advertised this from September 1904 until May 1908 including two periods when he said it was "on hire", and to add to the mystery, Messrs J. F. Jennings & Co. of Middlesbrough wrote to the L.B.&S.C.R. in January 1906 requesting sale of various spares for what appears to be the same locomotive.

K13 — Access to C. D. Phillips' advertisements between 1894 and 1903 reveals extra dimensions but no hard facts on disposals. The five small tank locomotives described in columns 1 and 2 were first advertised on 1st March 1894 "on completion of Alexandra Dock extension", when Mr. Phillips seems to have been acting as an agent for the A.D.R. as the locomotives were still in stock. The advertised prices were the same as those headed "cost" in the table on page K13 except that *Britannia* was rounded off to £480. This locomotive must have been bought by Griff Colliery right away as it does not appear in Phillips" advertisement of 2/4/94. He described it as being 0-6-0ST with 13 1/8" x 18" cylinders (fig. P50). The others continued to be advertised as below:

Usk was described as 0-4-0ST with 11 3/16" x 20" outside cylinders with 3' 6" wheels on a 6' 0" wheelbase. This rules out the B.&M. "little" *Usk* and lines 32-8 in Column 1, page K13 should be deleted. Mr. Phillips seems to have hired-out locomotives for short periods and an advertisement dated January 1898 said that *Usk* "can be seen at work". It was described as re-boilered in an October 1899 advertisement and as it was not advertised in November 1899, it presumably was sold in that period.

Ilkeston was described as 0-6-0ST with 11" x 18" inside cylinders and 3' 0" wheels on a 10' 3" wheelbase. It was advertised again in August 1895 but disappears in September 1896.

Harold has a complicated history; Kitsons' records of their No. 1829 of 1872 show it as a six-coupled tank locomotive with 13" x 20" cylinders and 3' 6" wheels, supplied so named to a contractor, T. B. Nelson, but with no location. It is possible, but by no means certain, that *Harold* was one of fifteen or more locomotives used by Nelson in the building of Cardiff's Roath Dock in 1883-87. Fifteen locomotives were offered at auction on 10/3/86, reduced to six in a later auction on 8/9/86 where they were described as four and six-coupled by Manning Wardle and Kitson. C. D. Phillips advertised *Harold* between 9/87 and 12/88, with the dimensions given on page K13, column 1, lines 53/4, so it must have been acquired by A.D.R. between 12/88 and 6/94. In his advertisements of 1/3/94 and 1/8/95 he quotes 13¼" x 20" cylinders with 3' 6" wheels (corresponding to the maker's figures) and on a 12' 3" wheelbase, 9" shorter than before. What appears to have been the same locomotive worked the first public passenger train on the Weston, Cleveland & Portishead Railway (see page K267) on 1st December 1897. A photograph taken on the W.C.&P. in 1897/8 show an inside-cylinder 0-6-0ST named *Harold* and numbered 45. Despite what was said on page K13, column 1, line 13, this leads to speculation that the five South Dock locomotives may have been numbered 41-45 in a separate A.D.R. series, or it could have been a C. D. Phillips number but the latter seems unlikely as *Harold* was described as *Emlyn No. 70* in his January 1899 advertisement. Phillips' lists describe it as "on hire" in February, July and August 1898 but as having been fitted with new cylinders in October. It was "on hire in Monmouthshire" in March 1899 and disappears from the list in May.

K14 — *Annie* was described in the March 1894 list as an 0-6-0ST by Peckett & Sons with 12⅜" x 20½" outside cylinders. The 7/99 list adds 3' 4" wheels on 9' 6" wheelbase, thoroughly overhauled, painted, lined and varnished. In the December 1902 and January 1903 advertisements it is described as *Emlyn No. 76* and does not appear in the December 1903 advertisement. There is no clue to its identity in the Peckett list and it probably was a locomotive by another maker rebuilt by Peckett with their "rebuild" plate attached. Delete the reference to *Annie* at the foot of column 1 and at the head of column 2, page K13.

K14 — Column 2, line 36: No. 679 was sent direct from Swindon to Johnston except for calling at Caerphilly en route for re-tubing and C. Williams seems only to have been an agent in the transaction. Hook Colliery passed to the National Coal Board and No. 10 was still at Johnston in August 1951 but had reached Trimsaran by 7/52. It was reported scrapped about 1956.

K15 — Column 2, line 55: Hunslet supplied their No. 281 to Isaac Llewellyn at Newport Docks, bearing the name *Admiral*. The name *Kate* seems to have been given by Messrs Dunn & Shute but as the transaction with the A. D. Co. mentions the name *Active*, Dunn & Shute probably changed the name during their ownership.

K16 — Column 1, line 13: Regarding the identity of *Alexandra*, on 10th April 1875 the first train of coal for shipment at Alexandra Docks was hauled by Monmouthshire Railways & Canals locomotive No. 44 to the junction with the A. D. Company's railway, to be taken over by the Dock Contractor's locomotive *Alexandra*. It this is the one which became G.W.R. No. 1341, it had been built by 1875, but there could have been two (or more) locomotives with such an appropriate name in the period 1875-1903. The locomotive which became G.W. No. 1341 could have been of Black, Hawthorn's make as the unusual 12" x 19" cylinders were a feature of this firm's standard 0-4-0ST. The tanks and the wheel centres are very much like those on a contemporary photograph of another Black Hawthorn locomotive but a Works List based on official data gives no clue to No. 1341's identity; certainly none of this standard type was supplied new to anywhere near Newport, nor to any of the contractors who worked at

LOCOMOTIVES SOLD OUT OF SERVICE — 1

Photograph] [F. Jones' Collection

Britannia **(ex A.D.&R.) at Griff Colliery, Bedworth.** P50

Photograph] [F. Jones

Ex-G.W. No. 711 at Bargoed in April 1956 with pannier tanks designed and fitted by National Coal Board. P51

Photograph] [F. Jones

Ex-G.W. No. 717 at Jingling Gate Colliery in 1958 with dual number plate reading 9 (20) P52

LOCOMOTIVES SOLD OUT OF SERVICE — 2
AND A MYSTERY PICTURE

Photograph] [W. Burkett

Ex-G.W. No. 154 as Hartley Main Collieries No. 27 with tanks as shortened in 1937. P53

Photograph] [H. C. Casserley

W. D. No. 101 (ex Rhymney Railway No. 024 or 026), Strawberry Hill shed, December 1921, en route from Bordon to Catterick. P54

Photograph] [P. J. T. Reed

An indistinct mystery photograph, Cardiff Cathays shed, October 1926, almost certainly of Taff Vale No. 84 as a stationary boiler. P55

Newport Docks.

K16 — Column 2, tables of dimensions: The Avonside Engine Co.'s list gives the cylinder dimensions as 14" x 20", so either the A. D. Co.'s drawing which formed the basis of Diagram K was wrong or the cylinders had been lined up to 12½" diameter. Diagram K was re-issued as Diagram P with weights, tank capacity and chimney height added, presumably by the G.W.R., but the cylinder dimensions remained at 12½" x 19¾". To add to the confusion, a second Diagram lettered "P" was issued to record the alterations made at Swindon in 1922/3, when either the cylinder dimensions were corrected or the liners taken out to restore them to the original 14" diameter.

K28 — Column 2, line 46: Evidence has come to light of Nos. 198/9, 206-8/10-12/14/23-6/30-2 having been fitted with vacuum brakes. Nos. 200/1/3/13/29 are known not to have been fitted and there is no information on Nos. 204/9/27/8.

K31 — Column 1, line 52: Nos. 238 and 246 were not overhauled by Kitson & Co., but by Kerr, Stuart & Co. of Stoke-on-Trent, as correctly shown on page 33, Part 1.

K31 — Column 2, last paragraph: Vacuum brake fittings were removed from most of the class (and some may have been withdrawn still so fitted) much later than 1932. Nos. 235/40-4/6-8/50/2-4/6-9/61-73/5-7 had them in July 1935 when a census was taken and several were seen with them up to 1938/39. Nos. 246/8/63/76/7 had lost theirs by 1947-50, Nos. 234/60 were not fitted in 1935 and No. 245 has the fittings on an early, undated photograph. Nos. 233/8/45/9/51/5/74 were not recorded.

K28 & K 31 — (see above): About one-half of the vacuum-fitted 0-6-2T had steam heating apparatus in the mid-1930's. Those without were probably intended for use on the large number of extra trains and excursions run to Barry Island at summer weekends and on Bank Holidays.

K41 — "F" Class Summary: No. 780 was rebuilt to Diagram B19 (Pannier Tanks) in April, not June 1927.

K41 — "F" Class Locomotives Sold:

No. 711, which retained its saddle tanks in March 1948, had been fitted with a National Coal Board design of pannier tanks by April 1954 (Fig. P51).

No. 713, as Bowes No. 8, is recorded as sold to D. S. Bowran on 1/4/46.

No. 714 had become Backworth Colliery No. 7 by April 1949.

No. 717 was photographed at Jingling Gate Colliery (the N.C.B. name for John Bowes' colliery) in 1958 with numberplates 9 (20) see Fig. P52. This was said to be a renumbering of No. 9 but the "20" obviously was the N.C.B. Area number as No. 13(20) had a similar dual number plate.

No. 718 (and No. 723, page K42): The amendment given on page M116, Part 12, was based on the Birmingham Locomotive Club's lists, of which an earlier one has been found, indicating that Nos. 718/23 exchanged numbers during a boiler change in 12/48 with the possible implication that these two as A.C.C. Nos. 22/3 changed to 23 and 22 in 1948 and reverted to 22/3 in 1957. There is no way of checking this but if true, No. 718 was broken up as A.C.C. 22 in 6/62 and 723 as A.C.C. 23 in 9/60.

No. 719 was No. 1 at the Rising Sun Colliery and became N.C.B. No. 10 some time between July 1949 and 1952. On 25/9/55 it was reported as "having been laid aside for some time" and was transferred to Backworth Colliery in November 1955. When its chassis was amalgamated with the boiler, tanks and cab of No. 714 (not the wheels of 714 as shown on page K41), it retained the number 10 and its Sharp Stewart makers' plates No. 4595 of 1900.

K42 — No. 780: There is confusion about the renumbering; in 1939 it was recorded at Nine Mile Point as No. A2 and on 5/2/41, Ocean Collieries, Treorchy (parent company of Burnyeat, Brown & Co.) advised "we have changed the number from 780 to A2", yet two later photographs show that the "780" plates were retained (with no sign of a number A2) and they were still in place in 1964. It was still at Nine Mile Point in September 1952 and at Arreal Griffin Colliery, Six Bells, Abertillery in August 1953 and June 1954. It was relieving No. 808, the regular locomotive at Blaendare Colliery, early in 1955 and was stored out of use at Hafodyrynys Colliery in December 1962.

K49 — Column 1, lines 4-7 to be replaced by: "G.W. 2-6-2T Nos. 3119/40 were on loan to Barry Shed from January 1920, to share these duties with the 0-6-4T (see page J28, Part 9). No. 3140 was replaced by No. 3129 in October 1920 and Nos. 3119/29 returned from loan in March and September 1921 respectively."

K53 — Special measures have to be taken when a railway is laid across a bog, like Whixall Moss, between Whitchurch and Ellesmere, and their effectiveness was tested in late April 1863 by running a train, described in a contemporary "Railway Times" account as having been hauled by "two heavy locomotives, *Montgomery* and *Hero*". *Montgomery* must surely have been Cambrian No. 5 (see page K56), brought by road as the Ellesmere-Oswestry line was not opened until July 1864 but there is no trace of either a Cambrian or a Savin locomotive named *Hero*. The most likely explanation is that the L.&N.W.R. "DX" Class 0-6-0 No. 192 *Hero*, built August 1860 was hired, as the new line *was* connected to the L.&N.W.R. Less likely, but still a possibility, is the St. Helens Railway No. 23 *Hero* which was in the same part of the country at the time. This locomotive later found its way to the Bristol Port Railway & Pier (see Part 3, page C81); if so, it had three extremely tenuous links with THE LOCOMOTIVES OF THE GREAT WESTERN RAILWAY as it also worked at a colliery connected with the Taff Vale Railway at Treorchy (see page M72, where the name *Dusty* should have read *Duty*).

K53 — Although the coast line from Dovey Junction to Pwllheli was not opened throughout until 10th October 1867, sections were opened in a disjointed fashion dictated mainly by the presence of unstable sandy foundations but in the case of Penmaenpool to Dolgellau, by delays in land acquisition and Parliamentary procedures. The first section, between Aberdovey and Llwyngwril, was opened on 24th October 1863, stock having been ferried across the Dovey Estuary from Ynyslas to Aberdovey. An extension from Llwyngril to Penmaenpool via what later became Barmouth Junction followed on 3rd July 1865. Although E. T. MacDermott's "History of the Great Western Railway" says that the Bala and Dolgelley Railway was opened throughout to a junction with the Cambrian Railways at Dolgelley on 4th August 1868, work on the Cambrian portion from Penmaenpool did not start until September 1868 and public traffic commenced to a temporary station on 21st June 1869. Running powers over the 28 chains of G.W. track into their Dolgelley Station were being negotiated and the first train to use the facility was the Sunday mail from Machynlleth on 1st August 1869. (In passing it may be mentioned that goods traffic, which normally preceded passenger trains, did not start running over the G.W. line from Bala to Dolgelley until 1st October 1868). Foundation difficulties at Dovey Junction had been overcome and enabled the line to be opened thence from Aberdovey on 14th August 1867; similarly completion of Barmouth Viaduct enabled the line from Barmouth Junction to Pwllheli to be opened, as stated, on 10th October, 1867. (The situation is complicated by the running of trains by the contractors before Board of Trade approval had been obtained but an advertised passenger service ran between Penrhyndeudraeth, Afonwen and Caernarfon with an Afonwen-Pwllheli connecting service as from 20th September 1867 and there was a horse-drawn service between Barmouth Junction and Barmouth from 3rd June 1867, possibly before the new viaduct had been passed for locomotives). The situation is further complicated in that the railway as far as Portmadoc was authorised by the Aberystwyth & Welsh Coast Railway Act of 22/7/61, with an extension to Pwllheli (and a further extension to Porth Dinlleyn on the north coast of the Lleyn Peninsula, which never was built) authorised by an Act of 29/7/62. The Carnarvonshire Railway's Act of 29/7/62 also authorised part of the same railway, from the junction at Afonwen back to Portmadoc. The conflict was resolved by an agreement of 13/12/65 under which what by now had become the Cambrian Railways built the Afonwen-Portmadoc section with running powers for the Carnarvonshire Railway. To add to the complications, the Carnarvonshire Company's locomotives were late arriving and two Cambrian locomotives were hired to run the service above described. The cross-country line became part of the L.&N.W.R. in November 1870.

K54 — Column 2, line 36: T. W. Worsdell's notebooks have been found and show that *Dwarf* was built in March 1862 for Messrs Savin & Ward; full dimensions are given, the principal being: cylinders 9" x 15" (not 14"), wheels 3' 0" on a 6' 0" wheelbase, length between buffer planks 16' 0", coal capacity 22.5 cu. ft., tank capacity 310 gallons, working pressure 100 lb/sq. in., weight empty 9 tons 18 cwt. Mr. Boulton resold to Messrs Barker & Co. of Wigan in April 1873.

K55 — Column 1, line 43: The makers did not fit nameplates to *Ruthin*, so the name may have been painted on which could account for its removal by 1866. The makers gave the length of the boiler barrel as 10' 3".

K56 — Column 1, line 11: *Nantclwyd* had a complicated history after its Cambrian days, which is not helped by conflicting accounts and what follows is a distillation of the evidence into the most likely story. Mr. Taylor was the contractor for the L.&S.W.R.'s Pirbright Junction to Farnham line (opened May 1870) and may have used the locomotive there but in 1870/1 the makers sent spares to the L.&N.W.R. There is no trace of a Taylor contract in the area to account for the locomotive being there but either then or later he lived at Whitmore, midway between Stafford and Crewe and could have been sub-contracting. The locomotive acquired the name *Whitmore* and there is a newspaper reference to its working on the Barnstaple & Ilfracombe Railway (built by Taylor in 1871-4) on 28/6/74. Taylor then built the Ascot & Aldershot Railway (1874-9) and spares were sent to him there by the makers, who later also supplied spares to George Furness of Higham, who built the Hoo Junction to Sharnal Street line (1880-82). A mysterious Mr. J. T. Chappell also was involved and seems to have paid for repairs to this locomotive by the L.B.&S.C.R. in 1875. While he is said to have effected the sale to the L.&S.W.R. in September 1883, someone unknown advertised a Manning Wardle 0-6-0ST named *Whitmore* in the C. D. Phillips Register between August 1883 and November 1886.

K56 — Column 1, line 21: The contractor for the Halesowen Branch and presumably the purchaser of *Cardigan* was Henry Lovatt.

K56 — Column 1, line 33: Manning Wardle supplied spares to the Llynvi Coal & Iron Co. in June 1870, to Jackson in about December 1871, to Walker at Shepton Mallet between March and August 1872 and to Docwra later. He was building the Canada Dock, Rotherhithe between September 1874 and November 1876, which fits the date of sale at Shepton Mallet in July 1874 (see page M116).

K56 — Column 1, line 44: *Lilleshall* is said to have been seen later at Brynkinalt Colliery, Chirk.

K57 &
K59 — It would seem from the tables on these two pages that the name *Countess Vane* was carried successively by Cambrian Nos. 35 and 41 but an official "List of Locomotive Engines 1866" shows Nos. 34 and 35 as *Countess Vane* and *Castell Deudraeth* respectively. They were referred to under their original names in the account of the official opening of the N.&M. on 3/1/63 and presumably still had these names when transferred to the O.&N. Joint Committee in May 1864. The report of an accident on the Carnarvonshire Railway at Brynkir in September 1866 said that the affected locomotive *Castell Deudraeth* had been on the line for ballasting work since May 1865. This does not prove that it was renamed at any time between January 1863 and September 1866 but there is a presumption in favour of May 1865 as it was named after the residence of the Chairman of the A.&W.C. which, as will be seen (addition to page K53 above) had close connections with the Carnarvonshire line. Although again speculation, the *Countess Vane* nameplates displaced from No. 35 would have been thought more appropriate than the place name *Talerddig* and the changes of name on Nos. 34 and 35 could have taken place at the same time.

Referring to footnote (b) on page K59, both the official 1866 list mentioned above and another of unknown origin but also dated 1866 show No. 41 still as *Cader Idris*. At some time from 1866 onwards Nos. 34 and 41 exchanged names so that *Countess Vane* was carried by three different locomotives in a fairly short space of time, ending up on No. 41. *Cader Idris* was successively No. 41 and No. 34; the name *Talerddig* removed from No. 34 was not used again until 1875, see page K63.

K65 — The Mid-Wales Railway had a L.&N.W.R. locomotive referred to as No. 2, *Hecla*, on loan in the 1880's. This number and name were used three times by the

L.&N.W.R. and the one still in service at the vital time was a "DX" Class 0-6-0 as it lasted until 1886. However, the "DX" locomotives lost their names in the late 1860's so unless someone with a very long memory revived the name, this locomotive doesn't fit. Another possibility — but no more — is an earlier No. 2 *Hecla* of 1844, sold in 1861 to Mercer & Evans who seemed to have owned some collieries south of Wigan and may have hired it to the Mid-Wales Railway.

K65 — Column 1, lines 23-6: Mid-Wales Nos. 2, 10 & 11 were built in 11/1864, 10/1865 and 10/1865 respectively; no month of building is known for No. 6.

K65 — Column 2, summary: No. 4 was built in 12/1864 and No. 3 (formerly No. 5) in 3/1865.

K70 — Column 2, line 27: No. 25 was sent to a shell filling depot at Credenhill, which was worked in conjunction with H. M. Filling Factory, Rotherwas, Hereford and to which No. 25 seems to have been transferred. It was advertised for sale at Rotherwas in November 1922 but did not appear in the Government "Surplus" list of January 1923. It could have been one of the two locomotives up for auction there on 14/12/22 and 21/3/23 or the single 12" standard gauge Manning Wardle 0-6-0 offered there on 30/11/23. There is no further trace.

K70 — Column 2, foot: An auction on 24/8/39 included an 0-4-0ST by Manning Wardle in 1901, obviously Cambrian No. 22, but no purchaser has been traced.

K72 — (facing): The original of illustration No. K88 shows the maker's number clearly as 1656, which was Cambrian No. 56, not 55 (see page K59) and the caption should be altered accordingly.

K77 — Column 2, line 53: No. 7 worked the last British Railways train over the Vale of Rheidol line on 4th November 1988 (fig. P4). The railway with locomotives and rolling stock was sold to the Brecon Mountain Railway Co. Ltd., of Pontsticill, Merthyr Tydfil, on 31st March 1989, on which date Nos. 7, 8 & 9 were taken out of British Rail stock, thus closing a chapter which began with the first Great Western passenger train on 31st May 1838 (see Part 2, page B4). More for convenience than logically, the three locomotives have been included in the "Preserved" lists in Parts 12 and 13; they are at last incontrovertibly in the right place in this Part!

K77 — Column 2, line 57: This change of livery was reported in *The Locomotive* for 15/9/15, which suggests a date some seven years after 1908.

K79 — Column 2, line 36: The reference should be to Part 3 (not Part 2) page C35 column 2, line 32.

K87 — Column 1, line 21 says there is no record of any locomotives owned by the Marquis before 1860; at the foot of column 2 there is reference to No. 1 (originally 3) and 4, built by Beyer Peacock (apparently as 0-4-2T) in 1860/61; while K90, column 1, line 7 refers to No. 24 as a Beyer Peacock 0-4-2T; K90, column 2, line 40 suggests that Nos. 2, 3, 6, 13 and probably No. 1 were not the first to carry these numbers. There is a temptation to look at these statements in relation to the Beyer Peacock list, which shows two locomotives supplied to "G. Thompson for Marquis of Bute". Maker's No. 45 was an inside-cylinder 0-4-2T, delivered or left works on 31/12/1856 and their No. 52 was an inside cylinder 0-4-0T, delivered or left works on 23/9/57.

K88 — Column 2, line 46: Fox, Walker's No. 200 (Cardiff No. 17) was at Brunner Mond, Winnington Works, Northwich, Cheshire, bearing the name *Cardiff* and was sold to an unknown buyer in 1930. Winnington Works seems to have been involved in work for the Ministry of Munitions, on whose behalf Brunner Mond also managed an explosives factory near Northwich and it seems highly likely that No. 17 went there in 1917.

K89 — Column 2, lines 23-6: the 0-6-0ST withdrawn in 1908 seem more likely to have been Nos. 14 and 16, because Nos. 20 and 22 were bought on 21st February 1917 by Thos. W. Ward Ltd, then lying at East Moors Locomotive Shed, whence they were despatched "to be scrapped" in May 1917 to Ward's yard at King's Dock, Swansea. Ward's quoted some dimensions at variance with those on Diagram A78 but as these included 12' 0" tubes in a 9' 5½" boiler barrel they are best ignored.

K92 — By 5/6/49, ex G.W. No. 692 was known as *Redbourne No. 38*.

K95 — Column 1, line 2: Hartley Main No. 27 appears with shorter tanks on a photograph taken in July 1937 (fig. P53), having been photographed earlier in

1937 with its original tanks.

K108 — Foot of column 1 and head of column 2: From what seem to be contemporary notes made by a railway enthusiast stationed at Catterick Camp during the Great War, W.D. Nos. 100/1 were Sharp, Stewart's Nos. 2234/6, establishing them as Rhymney Nos. 024 and 026, with W.D. No. 102 obviously Sharp Stewart's No. 2238 and Rhymney No. 028. The magazine *Locomotive News & Railway Notes* for 25/5/21 said "W.D. 101 was recently seen in London en route from Catterick to Bordon (for Longmoor)", yet the photograph (fig. P54) is captioned W.D. 101 "en route from Bordon to Catterick at Strawberry Hill, 31/12/21". It must have made the return journey to Longmoor again as it was clearly the one scrapped there in 1923. W.D. 102 was advertised in the Government publication "Surplus" as being at Catterick, from 1st May to 1st September 1922 and in another list, possibly an agent, from September 1922 to April 1923, but no sale has been traced. With Nos. 101/2 accounted for, the one sold to Workington Iron and Steel Co. must have been W.D. 100. Fig. P54 (Rhymney No. 026) shows an unmistakeable Rhymney double-framed 0-6-0ST which should finally dispose of the idea that W.D. Nos. 101/2 had been Brecon & Merthyr Nos. 4 & 5 (page K198). These started life as inside-framed 0-6-0 tender locomotives and may never even have been converted to 0-6-0ST.

K128 — Column 2, delete lines 15-21 and replace by: from March 1920, the three regular performers were Nos. 2731/43/54, which returned to the G.W.R. in March, April and April 1921 respectively. A Rhymney Railway incident book shows that Nos. 2734/47 replaced one or more of these in late 1920 but this obviously was a case of short-term replacement as they were allocated to Tondu and Cardiff respectively for the whole of the loan period.

K139 — *Llantwit* (see column 2) was one of two London & Croydon locomotives built by Rennie and shown in the South Eastern Railway history as sold on 12th April 1845.

K151 — Column 2, line 16: Although T.V. No. 84 is shown in the table as withdrawn in January 1902, a photograph taken at Cardiff Cathays on 31/10/26 (Fig. P55) shows what seems to be this locomotive, with coupling rods removed and cranks at varying angles, as if it had been used as a stationary boiler.

K171 — Column 1, line 45: The last hire period at Caerphilly Tar Plant occurred after official withdrawal of No. 194. It ceased work there on 14th April 1954 and was sent to Swindon on 26th April, not in May.

K178 — Column 2, *Radcliffe*: The T.V.R. advertised this locomotive on 7th October 1892 giving extra dimensions as wheelbase 10' 5½", boiler pressure 125 lb. and said that it could be inspected at Cathays Shed. The owners, W. R. Parker & Co. had filed a bankruptcy petition in connection with losses on a Rhondda Valley sewerage contract (published on 6th October) so that *Radcliffe* must have been in the area (perhaps on the sewerage contract) at the time. However, to complicate matters, Parker was awarded a contract in April 1888 for widening the Great Northern line between Saxondale and Radcliffe-on-Trent station which was completed in August 1890, with some ancillary work finished in November 1891. W. R. Parker could be the man who superintended the Penarth Dock contract for T. A. Walker; this was an extension to an existing dock, started in 1881 and completed in 1884. Thus the locomotive *might* have worked at Penarth, then on a Manchester Ship Canal contract (see page M118, Part 12), then in Nottinghamshire and finally in the Rhondda Valley.

K180 — Column 1, last line: In another case of the discrepancies mentioned under "General Notes" (above) a usually reliable source published in July 1958 says that N.C.B. No. 54 (ex-G.W. No. 475) was cut up in May 1958.

K183 — Column 1, line 19: No. 789 may not have been transferred to Varteg Colliery as a working locomotive. It was under repair at Blaenavon Ironworks in October 1947, reported "dumped" at Varteg in the summer of 1948 and derelict at Talywain in August 1949 with a condemned boiler (the three locations are only three miles apart).

K185 — Column 1, line 47: Although the "O4" Class were described as "mixed traffic" all batches had brass beading to the leading wheel splashers. This was removed from seven, and perhaps more, by G.W.R. and B.R.

K188 — Column 1, line 1: Makers' Nos. 7369-74 had been allotted by the time the order with the German firm had been cancelled.

K190 &

K192 — Summaries: The maker's plates on G.W. Nos. 398/9, 401 bore a 1920 date.

K199 — Column 2, line 31: C. D. Phillips advertised "*Little Usk*" as such between 1/9/81 and 1/5/82 but it was not in his 1/7/82 list. Amplifying the note on page M118, Part 12, it was working for Mackay on a Newport Dock contract in 1885.

K200 — Column 1, line 9: For "Coed-y-Go" read "Coed-y-Glo". As stated, Manning Wardle's No. 63 *Tiny* was sold to Mr. Savin in November 1868. In about 1873, the makers supplied spares to Chatterley Iron Co., and it may be more than coincidence that this Company sued Cambrian Railways for damage to a locomotive consigned from Fenns Bank (a Cambrian station near Whitchurch) to their Chell Colliery in August 1873. This colliery closed in 1910 and an unidentified 0-4-0ST turns up in the stock of Chatterley Whitfield Collieries Ltd., Whitfield, near Stoke-on-Trent. It was scrapped or sold at a later, unknown date.

K206 — Column 2, line 34: B.&M. No. 35 was advertised for sale by the Bute Works Supply Co. on 19/11/15 and 31/12/15.

K210 — Some published accounts show ex-G.W. No. 2161 as Ashington Colliery No. 3 but photographs taken in 1935 and 1949 both show it as No. 21.

K212 — The name of the valley was wrongly spelt "Gwendreath" in the title of the Company's Act of Parliament and the correct version is amply confirmed by maps and by the nameplates on G.W. No. 2196.

K212 — Column 2, line 15: The Contractor for the line was Frederick Furness, who could have been the original owner of *Lizzie* and *Gwendraeth.*

K212 — Foot of column 2: The series of trials on 20th June 1870 probably were arranged by Robert Fairlie as potential and actual customers from Canada, India and Mexico were present.

K213 — Column 2: delete lines 1-4 and replace by: . . Chatburn, as *Edith Mary*, in August 1874 for use on the construction of the Lancashire & Yorkshire's Chatburn-Hellifield railway. Barnes failed to complete the contract, which was taken over by T. J. Waller (whose interest in No. 459 was noted in Manning Wardle's records) and who completed the work in 1880. The next owner is reputed to have been the Bishwell Coal & Coke Co., of Swansea; it was then named *Susan.*

K213 — Column 2, line 12: G. K, Waghorn was a contractor and on 15/4/05 a 13" 0-6-0ST by Manning Wardle was offered for sale on his behalf. The Manning Wardle entries about his ownership seem to cover the years 1901-5; in their records the name is spelt *Copshawe.* He is believed to have owned only one locomotive.

K214 — Column 1, line 14: *Victoria* was broken up at Burry Port in 1903, presumably by the unknown buyer as the price of £200 suggests scrap value.

K216 — Table of Locomotives and K217, column 2, line 23: No. 13, which left Hudswell Clarke's works on 20/10/16 also was commandeered by the War Office and sent to Richborough Sidings, Sandwich, Kent and the makers made arrangements to supply another identical locomotive, their No. 1288, as "B.P.G.V. No. 13". In the event, the intended No. 13 (maker's No. 1222) was released to the B.P.G.V. and "No. 1288" went to the Inland Waterways Depot at Chepstow, ex works on 31/8/17.

K218 — Column 1, line 47 (and Part 13, page N34): Messrs Bott & Stennett were closely involved, as contractors for building the railway and promoters of the Abdon Clee Stone Quarry Company; Mr. Stennett was also a Director of the C.M.&D.P. Pending delivery of the Company's locomotives (which started trial trips on 20th August 1908), the contractors worked the freight traffic for 50% of the gross receipts. There is photographic evidence that 0-4-0ST *Canada* and 0-6-0ST *Uxbridge* shared the work with *Fleetwood* (which was not "hired" as stated on page N34). These were Hunslet Engine Co.'s Nos. 525 of 1890 and 761 of 1902 respectively and were not transferred to the Quarry Company like *Fleetwood,* but disposed of elsewhere. The passenger service started in November 1908 (not 1909 as given in line 39, page K218).

K219 — Column 1, line 22: No. 28 was further Great-Westernised before fitting of pannier tanks, acquiring a G.W. safety valve and other features in 1925/6.

K219 — Column 1, line 26: No. 29 was fitted for steam heating in 1924 and two ex-

C.M.&D.P. carriages were similarly fitted at the same time, a welcome improvement on the old days!

K220 — Column 1, lines 17 and 21: What was believed to have been a Fox Walker official list contains at least some informal additions which may have led to confusion between Fox Walker's No. 150 and Manning Wardle's No. 150. The latter was supplied new to Dowlais Ironworks bearing the name *Tir-y-colly*, which may have no relevance to the account on page K220.

K220 — Column 2, line 19: The Anglo-Scottish Construction Co. completed construction of the Trienta y Tres to Rio Branco section of the Uruguay State Railways in 1935. G.W. No. 26 presumably was used on this contract and as the railway was of 4' $8^{1/2}$" gauge, could well have been retained for further use.

K221 — Column 2, line 34: Andrew Barclay's No. 221, which was 0-6-0ST, (Fig. P56) was advertised by the dealer, J. F. Wake, between 1/12/11 and 17/7/14. He quotes a wheelbase of 11' 0" and a length over buffers of 25' 9", but yet another version of the wheel diameter at 3' 8"; since versions of wheel diameters could vary with new tyres and even inaccurate measuring, there is no point in trying to reconcile the alternatives of 3' 6", 3' 8" and 3' 9".

K221 — Column 2, line 43: There is no trace of a Waddell contract on Tyneside in 1885 and this reference is best ignored. The most likely version, which must still be accepted with reserve, is that Fox Walker's Nos. 278/9 of 1875 were supplied to W. F. Lawrence, the contractor for the sections of the Banbury & Cheltenham Direct Railway between Kings Sutton and Chipping Norton and between Bourton-on-the-Water and Cheltenham. F.W.'s No. 278 was named *Earl of Devon* and as he and *Seymour Clarke* were respectively Chairman and Vice Chairman of this Company, it suggests that the two locomotives went to this contract . Lawrence gave up the contract late in 1876 or early in 1877 and the next reference to No. 279 in Fox Walker's records is that it was with Waddell in Scarborough. Waddell's built the Scarborough and Whitby Railway between 1882 and 1885 but might just have used the locomotive on construction of the contiguous Whitby-Loftus Railway which they also built, between 1878 and 1883.

Although no L.&M.M. connection as such has been found, another Fox Walker 13" cylinder 0-6-0ST, their No. 264 of 1875 has been reported as with "John Waddell, Contractor, Llanelly" and with "Lawrence" which suggests at least an informal link between the firms, assuming that the same Lawrence was involved.

K222 — Column 1, line 20: The accident report indicates that *Inveravon* arrived in Llanelly in 1887, yet Waddell had ordered spares from Andrew Barclay in March 1885, indicating purchase from Scott & Best by then but there is no evidence that he used it in Scotland.

K223 — Column 2, line 56: No. 803 carried the temporary "W" suffix in 2/48 and still had it in 7/50. Delete the reference to the smokebox number plate.

K224 — Column 1, line 6: *John Waddell* was not dismantled at Great Mountain Colliery. It was transferred to Morlais Colliery, Llangennech about 1958 and was broken up there in about September 1964.

K225 — The L.&M.M. did not provide passenger trains but *Victory* (later G.W. No. 704) was fitted with the vacuum brake as built (and still retained it in 1937) and there is a possible connection with the experimental steam railmotor service of May 1920, described on page N16, Part 13. The "Uncoloured, A" classification given on page K225 seems to have been as recorded at Swindon but No. 704 carried an "A" on a yellow disc and one or more of the axle loads could have exceeded 14 tons as the C.M.&D.P. locomotives, also built by Manning Wardle, had much less weight on the trailing coupled wheels (Page K219). No. 803 (see page K223, column 2, line 49) was fitted with vacuum brake and steam heating at Swindon in January 1931 for passenger work on the C.M.&D.P.L.R.

K227 — Column 2, last paragraph: The M.&S.W.J. actually purchased thirty-one locomotives but the extra one did not appear in stock returns. This was an 0-4-0ST with 10" x 16" outside cylinders, Manning Wardle's No. 1259 of 1894, supplied new to Cardiff Corporation, Merthyr Tydfil, for reservoir construction and named *Beacon*. It was sold in September 1898, presumably to Bute Works Supply Co., who resold it to the Company in November 1898. The makers supplied spares to the Company and *The Locomotive* for 15/3/1910 reported it shunting the Cirencester Locomotive Shops. It was not in stock when the G.W.R. took over in 1923 and had been reported at the Blaenclydach Colliery, where it was said to have been cut up in July 1928. (The actual purchaser in 1898 seems to have been the Marlborough & Grafton Railway, presumably a M.&S.W.J. subsidiary, which may account for its not being taken into M.&S.W.J. stock).

K229 — Column 2, line 37: M.&S.W.J. board minutes show that No. 14 was withdrawn from active service at the end of April 1913. It was advertised by the Bute Works Supply Co. on 15/8/13 and the minutes record a sale to J. F. Wake in March 1914. It was despatched to him by G.W.R. on 3/6/14 and a new boiler, cab and tender are reputed to have been provided by the North British Railway at Cowlairs Works "about 1915" and it is said to have been stored during the War years at two Ministry of Munitions depots in Scotland. J. F. Wake advertised a 2-6-0 locomotive by Beyer Peacock, with six-wheel tender, 18" x 26" cylinders and 4' 0" wheels in October 1918.

K235 — Cancel the "correction" on page N34, Part 13. The train *did* go through to Brecon but as the line had not been passed by the Board of Trade, the passengers — 700 of them according to a contemporary newspaper — were conveyed free, but the trip was not such a bargain as it seemed. The Brecon weather was atrocious and the return train, due to leave at 8.30p.m., started two hours late and took a further seven hours to get back to Neath.

K236 — *Progress* took part in trials on the Mid-Wales Railway at Three Cocks Junction on 14th February 1870 in the presence of railway officials from France, Germany, Norway, Russia and Sweden, apparently invited by Robert Fairlie.

K237 — Column 2, line 20: Hiring from the Mid-Wales Railway may be connected with a reported through working between Llanidloes and Neath in the 1870's, when a Mid-Wales locomotive worked to Neath on Mondays, Wednesdays and Fridays and returned to Llanidloes on Tuesdays, Thursdays and Saturdays.

K237 — Column 2, line 23: The details of *Bulkeley* (or *Buckley*) given here are from the particulars of the same auction in November 1868, when *Progress* also was offered for sale. (Page K231, column 1, line 48). The auctioneer seems to have confused the details with those of an unidentifiable locomotive named *Neath* which he ascribed to Beyer Peacock.

In June 1866, Beyer Peacock arranged to supply parts for *Bulkley* (spelt thus) to the order of the N.&B.R. The order was cross-referenced to Beyer Peacock's Nos. 91 and 92, 0-4-0ST with 11" x 16" cylinders and 3' 0" wheels, built in 8/1858 and 9/1858 for the Broughton Coal Co. and the Brymbo Co. respectively and "Bulkeley" (however spelt!) could have been one of these. Mr. Forwood might already have owned it, or may have bought it at the auction as a locomotive of this name was hired from him in the latter part of 1870, see page K237 as above.

K243 — Column 1, line 23: P.T.R. *Derby* went to Swansea Docks between June and December 1897 and not "between 1895 and 1899". The correct version appears on page 257, column 2. Although Swansea Harbour Trust only kept this locomotive for about a year, they had advised Peckett & Sons of their interest.

K243 — Column 1, line 38: J. E. Billups built a section of the Taff Vale Railway from Roath to Cardiff East Moors, passing through the Penylan area of Cardiff (hence the name), between 1886 and 1888. A Railway Clearing House book of about 1890 lists a Billups' siding at Gellyrhaidd Junction, Tonyrefail on the Llantrisant to Penygraig branch but no new public railways were built here at the time. He offered three six-wheel coupled tank locomotives for sale at his Cathays, Cardiff yard on 3rd March 1892. Little is known of the Cefn & Pyle Railway, but in October 1892, the company advertised "a second-hand four-wheel tank locomotive with 10" x 15" outside cylinders, 2' 8½" wheels on a 5' 0" wheelbase which could be seen at their locomotive shed near Bryndu, near the G.W. station at Pyle". This just might signify the end of a construction project with the need for larger locomotives to work the traffic. Finally, an entry in Manning Wardle's list, undated, but probably in the 1892-4 period, shows their No. 955 with the Cefn & Pyle Railway.

K247 — Column 1, line 18: No. 812 was awaiting scrap at Varteg Colliery in August 1948 and had been scrapped by 5/6/49. Delete "Scrapped 1950".

K248 — Column 2, line 35: L.&N.W.R. official records show the sale to the R.&S.B. on 10/7/84, which corresponds to the into-stock date of "the latter half of 1884" but it was not despatched to Swansea until 10/2/85.

K254 — Column 1, line 17: The line was leased to the Glyncorrwg Coal Co. by the South Wales Mineral Railway (Lease) Act of 25/5/55, for 30 years. The Coal Co. went bankrupt in 1869 and the line seems to have been worked by a Receiver. Fresh agreements for working the line were made (with what had become the Glyncorrwg Colliery Company Ltd.) as from March and November 1880. Minutes of a S.W.M.R. Board Meeting on 13/4/72 said "the four broad gauge engines belonging to the Railway Company but leased to the Colliery Company should be sold and the money divided by an Actuary between the Railway and the Colliery Company. The change from broad to narrow gauge would have to be borne by the Company, the Colliery Company having to find new engines." G.W. records show all six locomotives as supplied to the Colliery Company, in 1872, 1875 and 1905, yet S.W.M. Reports & Accounts for years ending December 1868 to December 1877 show three locomotives, with a "nil" return from December 1878. Purely on a statistical basis, the "four broad gauge engines" seem to have been *Princess, Glyncorrwg, Brigand* and the unidentified locomotive.

K254 — Column 1, line 49: Manning Wardle's Nos. 135 and 137 were both dated 1864 and as their records show No. 136 as delivered in February 1866, this might indicate that the order had been placed by the S.W.M.R. and cancelled.

K254 — Column 2, lines 23-8: A locomotive with 0-4-2T wheel arrangement with 14" cylinders, named *Princess*, was offered for sale on 21/1/02 by C.E.&H.M. Peel of Swansea. The firm is described as Consulting Mechanical & Electrical Engineers in advertisements in Machinery Mart between 3/02 and 2/03, where what seems to be the same locomotive is described as having six wheels. Until 7/4/02, Landore Works was owned by Messrs Wright, Butler & Co., so that if the hiring did take place to their successors Baldwins Ltd., this would have been in April 1902 or later. Incidentally, a scrap merchant named Peel of Swansea bought a P.&M. locomotive within the next ten years or so — see page K261).

K255 — Two photographs of the Wolverhampton-built 0-6-0ST in S.W.M. service showing alterations appear as figs. P57 and P58.

K255 — Column1, line 36: According to Black, Hawthorn records, No. 5 was ordered on 23/10/90 "for delivery in eighteen weeks" so the Swindon date of 1890 is clearly wrong. The original of fig. P59 shows the 1891 date on the maker's plate.

K256 — Column 2, line 40: The Co-operative Wholesale Society had a preserves factory at Middleton Junction, about five miles from Manchester, and *Jumbo* may have gone there. Also, in June 1891 Mr. Westlake advertised two four-coupled tank locomotives, one with 10" and one with 12" cylinders.

K256 — Column 2, line 45: *Glanmor* was under repair at the Nevill's Dock Co.'s Llanelly workshops in 3/48 and was acquired later by Cudworth & Johnson of Wrexham, who rebuilt it in 1949. Hired by them for use at Messrs Rea's Ore & Grain Quay at Birkenhead Docks in 10/49, it seems to have been on hire at the Docks for the next twenty years or so. It was photographed there, still named *Glanmor*, in 4/60 (Fig. P60). It was noted laid aside with other locomotives on 17/11/62 and had disappeared by 29/2/64, stated to have been scrapped on site.

K257 — Column 1, lines 8-12: Delete "It is reported . . . dock shunting". Mr. Westlake, who also appears to have traded as the Low Level Haulage Company of Swansea, advertised this 0-6-0 for sale in June 1891. The Mersey Engineering Co. of Swansea Docks advertised what must have been the same locomotive in November 1894 because in both cases the wheelbase was given as 14' 8" albeit with two more variations from the cylinder dimensions given on page K248. A "secondhand R.&S.B. locomotive" was hired to the Cambrian Railways from mid-August to the end of September 1892 to make good a temporary shortage caused by a collision on the Talerddig Incline; the Cambrian declined an offer to buy it for £625.

K257 — Column 2, line 12: The Castner Kellner factory was at Runcorn.

K257 — Column 2, line 17: "R5" must have been returned from loan as it is said to have been acquired in 1937 by John Cashmore Ltd when they dismantled the Tipton furnaces and probably scrapped the locomotives then on hand.

K257 — Column 2, line 24: No. 4 (and No. 5, see line 52) *were* sold to C. D. Phillips and advertised by him from October 1911 onward; he succeeded in selling No. 4 to Billingsley Colliery on 11/3/13 but ceased to advertise No. 5 after June 1912 and probably scrapped it himself or sold it for scrap (FIg. P61).

K258 — Column 1, lines 32 to 42: Nos. 1 and 2 both were sold back to Hudswell, Clarke whose records have no reference to either the Ministry of Munitions or Williams, Foster & Co. and these eleven lines should be deleted and replaced by: Hudswell, Clarke's records show that No. 1 was supplied to Graigola Merthyr Colliery in March 1915 (and probably did not go to their works, but direct to the colliery). It was still there when vested in the N.C.B. in January 1947, was transferred to Felin Fran Colliery in March 1956 (or 1957, two versions differ) and was scrapped in about 1966. Hudswell, Clarke bought No. 2 on 1st August 1915 and up to February 1924 had hired it to eleven firms. In June 1924 it was sold to Semet-Solvey & Piette Coke Oven Co. Ltd., and despatched to South Yorkshire Chemical Works Ltd., Parkgate, Rotherham, presumably to work on a contract there as they advertised it for sale on 16th June 1926. It may have been taken over by South Yorkshire Chemicals and is variously given as scrapped in about 1939 and in about 1949.

K258 — Column 2, line 43: No picture of G.W. No. 933 has been found but its varied earlier and later history is illustrated and described in Figs. P62/3/4.

K261 — Column 1, line 22: No David Hinds has been traced at Cross Hands but someone of this name owned Llanmorlais Colliery, Penclawdd, near Gowerton in 1914.

K261 — Column 1, line 33: All the locomotives built by Markham & Co. Ltd., Broad Oak Ironworks, Chesterfield are too well-documented to be possibilities, but it is possible that P.&M. No. 2 was built by the Butterley Co. Ltd., Butterley, the only other known locomotive builder in the Chesterfield area. (The Staveley Coal & Iron Co., who also had works in the area, are reputed to have built some locomotives but they do not appear in any list of locomotive builders). Yet another possibility is a Vulcan Foundry 6-wheeled locomotive with 12" x 17" cylinders and 2' 10" driving wheels, advertised by the Bute Works Supply Co. on 27/7/00 as "now at Chesterfield".

K261 — Column 2, line 41: "795" was last reported at work on 18/1/57 and in October 1962, reported to have been "withdrawn some years ago and now lying behind the shed stripped of brasses and cab fittings".

K262 — Column 1, line 28: ex-P.&M. No. 10 almost certainly worked at the Heath Town phosphorous factory, Wolverhampton, built for the Ministry of Munitions and operated on the Ministry's behalf by Messrs Albright & Wilson. Standard gauge sidings were installed in 1917 and the works were offered for sale from mid-1920 until at least July 1923 but there is no mention of locomotives as such.

K263 — Column 2, line 39: No. 1151 had the number "1143" crudely painted on at R.S. Hayes' depot and had been mistaken for 1143 as the previous number of 1143 (968 — see page K260, top of column 2) was stamped on the motion. This arose because when 1143 was being cut up at Caerphilly in late 1960, its wheels and motion were transferred to No. 1151.

K264 — Column 1, line 58: The version of maker's numbers 324/2/3 on this line was taken from the maker's plates on 9/6/25, but the date of 1878 on the plates shows that they had been built by Henry Hughes & Co., whose business was not taken over by the Falcon Engine & Car Co. Ltd. until 1880. Further research shows that the entry on page M119, Part 12, can be discounted; in particular, a locomotive at Devonport Dockyard built by Henry Hughes in 1880 bore the maker's number 340. The address of both firms was Falcon Engine Works, Loughborough, which could have been inscribed on the 1878 plates.

K267 — Column 1, line 6: No. 5 retained its nameplates but was fitted with a G.W. whistle instead of the Stroudley type.

K267 — Column 2, line 4: For further details of *Harold* see page P64.

K267 — Column 2: Replace lines 28-30 by "*Walton Park*, Hudswell, Clarke No. 823 of 1908, bought new and transferred in 1913 to the Shropshire & Montgomeryshire Railway, also managed by Colonel Stephens and moved again in August 1916 to the East Kent Railway (also managed by Col. Stephens), where it retained its name but also became E.K.R. No. 2. It was sold to T. W.

Ward in 10/43, who resold to the Purfleet Deep Water Wharf & Storage Co, at whose premises it was seen, named *Churchill*, in 9/46. It remained there until scrapped by George Cohen in 7/57".

K269 — Column 2, line 41: Thirty-three of these were made fit for service at Oxley shed and the remainder at various depots throughout the system. Those dealt with at Oxley had been stored at Royds Green, the rest at Stratton.

PART 11 — THE RAILMOTOR VEHICLES AND INTERNAL COMBUSTION LOCOMOTIVES

Page

L6 — Column 1, line 45 of First Edition, column 1, line 47 of Second Edition: Delete "evidently". Nos. 15 and 16 were of Kerr, Stuart's design and the engine units were built at their Stoke-on-Trent Works. They were sent to Bristol for fitting into the carriage bodies built by the Bristol Wagon & Carriage Works Co. Ltd. (see reference to page L4 on page M125, Part 12).

L8 — Both Editions, page L8: Although Cars Nos. 59 and 60 were reputed to have been built with vestibule connections, these were not shown on Diagram P but did appear, in exaggerated form, on Diagram T.

L9 & L10 — Both Editions, Tables 3 and 4: Although G.W. records show that Cars Nos. 42 and 49 were withdrawn in July 1920 together with Engines Nos. 0807 and 0855 respectively and sold on 10th September 1920, the records also show Engine No. 0825 as withdrawn in October 1922, also in Car No. 42. An article in *The Industrial Locomotive,* Autumn 1984 issue, says that two railmotors were bought by the Port of London Authority for the Millwall Extension Railway in 1922,not 1920 and that P.L.A. Car No. 1 (ex G.W. No. 42) "had only one trial run after which it was laid aside and never used again". The article gave the engine numbers as 0825 of 1904 and 0888 of 1905, presumably from the numbers on the frames. "0888" had clearly been misread for 0858, as 0858 *was* built in 1905, while 0888 built in 1907 lasted in G.W. stock until December 1934.

It would be unwise to draw firm conclusions from these opposing accounts but it seems likely that Engine No. 0807 failed in the trials referred to and that 0825 replaced it in 1922. Perhaps neither of the railmotors was used in regular service until 1922, which could account for the P.L.A. "into stock" date. It certainly seems unlikely that Car No. 1 with the replacement engine 0825 would have continued to have been laid aside.

L12 — Column 1, line 52, First Edition, Column 2, line 12, Second Edition: The Glasgow Railway & Engineering Co. Ltd., Govan did not close until 1959 but their records have not yet been located.

L12 — Column 2, line 53 of First Edition, page L13, column 1, line 19 of Second Edition: It is strange that Steam Railmotors disappeared from the annual returns in 1917 because Car No. 3 was photographed at work on 20th May 1919. Also, a stock register found at Swindon said that one steam car had been taken over from the Cardiff Railway in May 1922. It may by then have ceased work and have been awaiting conversion, or was being converted, to a trailer. Until the Cardiff Railway bought four carriages from the Hull & Barnsley Railway in April 1919, the only passenger stock comprised two railmotors and two trailers (see reference to use of trailers with 2-4-2T No. 36 on page K95, Part 10) and it seems unlikely that the other railmotor (No. 2) would have been released for conversion until after April 1919. Both conversions are reputed to have been done by the original builders, the Gloucester Railway Carriage & Wagon Co. Ltd.

L13 — Both Editions, Port Talbot Railway & Docks Co's No. 1: Although built in 1906 it did not enter service until 10th May 1907 and spent the equivalent of a year and a half in Swindon Works between 1908 and 1913, spread over five visits. It finally left Port Talbot on 7th August 1915 and was stored in Swindon Stock Shed from 30th October 1915 until formal withdrawal in July 1920, being sold to the Port of London Authority on 10th September 1920. An accident report says that "SRM No. 1" was involved in an accident when preparing to work the 5.45 a.m. to Blaengarw so it certainly worked on the P.T.R. main line.

L16 — Both Editions, Auto Train Working: A general picture of the situation in 1911 is given in Table 2A (earlier) as amplified in the notes preceding the Table.

L17 — Column 2, line 46 of First Edition, page L18, column 2, line 43 of Second Edition: With the sale of ex-Vale of Rheidol 2-6-2T Nos. 7, 8 and 9, the only survivor of the subjects of this series in British Rail stock is the former Auto Trailer No. 233, later B.R. No. W233. It was converted to Departmental use in 1966 and is known as TEST CAR 1. It retains its former "drive end" and is in chocolate and cream livery. Now known as Departmental No. ADW 150375, it is currently used for the monitoring of ride tests on freight vehicles. In 1982 it was re-bogied and the original bogies sold to the Great Western Society, Didcot. It is normally kept at Derby.

L18 &
L19 — Column 1, line 14, on page L18, First Edition and L19, Second Edition: Diagram "U" shows the wheelbase as 19' 0" and the length over buffers (not shown on Diagram "U") was given officially as 36' 7".

L18 to L21, First Edition, L19-L22, Second Edition: An official record has been found giving technical details. The vacuum brakes on Cars Nos. 1 to 18 were an innovation in that they operated through brake drums, but there are no known reports of problems with leaves on the line; however, Cars Nos. 19 to 38, whose underframes were built at Swindon, had conventional clasp-type brakes.

L21 — Second Edition, column 1, line 14: For 24th March 1954 read 24th March 1951.

L21 — Second Edition, column 2: Railcars Nos. 1-16 could not be worked in multiple and their limited seating capacity had a bearing on their duties but, with typical Great Western enterprise, they entered the charter market. A Bristol car was chartered to take a party of bell ringers to Newquay, a round trip of over 360 miles while a Gloucester car (No. 7) was chartered for a comprehensive railtour of the remaining lines in the Forest of Dean network in September 1950. They have been used by the Engineers for Inspection Specials and a Pontypool Road car took part in an experimental increase in service frequency on the Pontypool Road to Monmouth branch which, sadly, did not create enough extra traffic to justify retention. A Landore car worked Saturday afternoon pre-booked excursions to Tenby and these few examples will serve to illustrate the versatility of the design.

Another even more enterprising venture, sadly postponed on the outbreak of World War 2, was the proposal to build a cut-off line from St. Germans to Looe, with the intention of operating a half-hourly through service between Plymouth and Looe, to be worked by single railcars.

L22 — First Edition, L23 Second Edition, both column 1: A small internal-combustion locomotive, about which only its number — P.W.M. 1024 — is known was built in 1930 by R. A. Lister of Dursley for use at Swansea Docks. It was broken up in April 1951.

L24 — First Edition, L25 Second Edition, both references to Diagrams of Gas Turbine Locomotives: Diagram F and G gave no details of the Power Groups to which these unusual locomotives belonged, nor of the Route Colours defining where they were allowed to run.

PART 12 — A CHRONOLOGICAL AND STATISTICAL SURVEY

Page
M11 — Column 2, lines 20/21: Delete "and even an occasional Bulldog" and replace by: "and with "Bulldog" No. 3434, five "Cities", ten "Counties" and nine of the 4100-68 series".

M14 — Column 1, lines 17/8: Four tenders, Nos. 4011/2/7/9 were welded.

M17 — Column 2, line 38: Swindon's workload was gradually run down and final closure was announced in 1985 and at a most inopportune time as preparations for the G.W.R.'s 150th anniversary were in hand. Not unexpectedly, nobody at Swindon wanted to celebrate and some of the more interesting planned events were cancelled. The site was sold for redevelopment and the western half

ABSORBED COMPANIES' LOCOMOTIVES — 1

Photograph] [F. Jones' Collection

This faded picture is of Andrew Barclay's No. 221, formerly L.&M.M. *John Waddell*, **at J. F. Wake's Darlington yard.** P56

Photograph] [R. Simmonds' Collection

The original of this S.W.M. picture shows a figure "1" on the cab side. This would have been Wolverhampton Works No. 190 of 1872 and is here shown as built except for the improved cab. P57

Photograph] [W. Watkins

Another Wolverhampton-built S.W.M. 0-6-0ST (one of Nos. 2-4) with a different style of overall cab, otherwise as built. P58

ABSORBED COMPANIES' LOCOMOTIVES — 2

Photograph] [Allan Baker's Collection
Black Hawthorn's official photograph of S.W.M. No. 5. P59

Photograph] [F. Jones' Collection
Ex-Wm. Westlake's No. 1 as *Glanmor* **at Birkenhead Docks, April 1960 — 75 years young!** P60

Photograph] [F. Jones' Collection
C. Rowland's No. 5 at Swansea Docks. The plate reads "R No. 5"; the picture does not show any name. P61

EVERYTHING BUT G.W. No. 933!

Photograph] [National Railway Museum

C. Rowland's No. 10 at Swansea Docks 21/7/04. P62

Photograph] [F. Jones' Collection

Ex-G.W. No. 933 as restored by A. R. Adams & Son. P63

Photograph] [F. Jones' Collection

Ex-G.W. No. 933 as ***John*** **at Bomarsund Colliery, 22/4/49.** P64

EXCEPTIONS TO THE RULE

Photograph] [National Railway Museum

0-6-0ST No. 1217 showing "PN" shed code in brass letters on the cab side, just below the roof. P65

Photograph] [B. Matthews' Collection

No. 2146 at Gloucester 13/5/50, the only known example of the class with steps on the fireman's side of the bunker. P66

Photograph] [C. F. H. Oldham

No. 7006 at Stratford-upon-Avon 2/10/48 showing the inverted "U" shape pressing covering an anti-carbonising valve (see Part 12, page M98). P67

including the celebrated "A" Shop, has mostly been cleared of structures. Some of the remaining buildings are being considered for incorporation in a Heritage Museum and proposals are being worked out at the time of writing. In the meantime, part of the former "R" and "B" Shops (incorporating the early broad gauge shed) has been leased to "Swindon Workshops Ltd".

M18 — Column 2: Add to the last paragraph under "Caerphilly": The works has been converted to industrial use and part of it is occupied by the Caerphilly Railway Society.

M18 — Of the factories listed under the heading "Others", Newton Abbot is intact but out of use, at Barry only the Boiler House remains, and Oswestry has been converted to industrial use.

M20 — Mr. R. L. Pittard has provided details of all the former Great Western steam depots still known to exist, albeit for other purposes. These are dealt with below but recent research reveals that four early sheds not mentioned in Part 12 still stand: Bristol South Wales Junction (closed 1876/7) is still used for wagon repairs 115 years later; the broad gauge shed at High Wycombe is used as a warehouse; the original South Wales Railway shed at Swansea High Street had been incorporated in the Goods Station which in turn has been closed and is used as a warehouse; and the original Swindon shed, which was incorporated in extensions to Swindon Works and survives as part of the premises used by "Swindon Workshops Ltd.", see above.

M21 — Column 2, line 23: The shed at Newport High Street, closed in the 1900-15 period, is still in existence as a garage.

M23 — Column 1, line 52: The shed codes PN (not PDN), SHL, SLO and RDG (but apparently no others) were shown by brass letters attached to the cabsides of locomotives stationed there (Fig. P65). They are known to have been in use in the 1920-24 period; 0-6-0 No. 2572 was photographed thus on 20/8/21 and still carried the "RDG" letters at Machynlleth in 1943!

M24 — Column 2, table: There was a second shed at Rosebush, in the quarry, closed at an unknown date but the side walls still stand.

M26 — Column 1: Of the sheds closed in the 1922-35 period, the following still exist in the form shown: Radstock (rail-connected and used by a preservation society); Shipston-on-Stour, used until very recently as a light engineering workshop; Newport Dock Street and Dinas Mawddwy (both industrial); Kerry (used as a barn!); Cheltenham High Street (Warehouse). Taffs Well, see column 2 (Garage). Two pre-Grouping survivors, not listed in Part 12, are the A.D.R. shed at Newport Dock (closed on opening of Pill Shed in 1898), used as a docks workshop; and the Van Railway shed at Caersws (closed by the Cambrian Railway between 1912 and 1922) which is a small factory.

M27 to
M30 — The "B.R. 1950 Code" forming a column heading on the left hand side of each of these pages applied until 1963, when there was a wholesale exchange of locomotives and of locomotive depots with the London Midland and the Southern Regions. These changes, however logical geographically, destroyed the identity of the Great Western Railway. The situation was complicated by shed closures and by the transfer of some sheds between Divisions on the Western Region and the altered shed codes are not relevant to a G.W.R. history. They were detailed in the Railway Observer at the time, on page 311 in October, page 342 in November and page 386 in December 1963.

All the structures have been demolished at over two-thirds of the sheds listed on these pages and some sites remain vacant while others have been redeveloped. Most of the remaining depots have been put to alternative uses as listed here in Divisional order:

London Division: The repair shop at Old Oak Common is used for maintaining diesel stock and the offices are still occupied. Southall shed is used both by the G.W.R. Preservation Group and also for stabling and servicing of main line steam locomotives used on special trains organised by Flying Scotsman Services Ltd. and the Steam Locomotive Operators' Association; the turntables here and at Old Oak Common are still in place. Reading has been completely rebuilt as a diesel maintenance depot but the pumphouse and water-softener store remain; one shed wall remains at Henley; Didcot has been extended by the Great Western Society but the original structures are intact.

Bristol Division: The repair shop at Bath Road is used for HST maintenance and the turntable still exists. Malmesbury is in industrial use.

Newton Abbot Division: Newton Abbot shed is intact but unused; Ashburton is in industrial use; Moretonhampstead is part of a transport depot; Taunton Repair Shop is rail-connected and used for maintaining track machines; Bridgwater is in industrial use; Exeter shed offices are all that remain and are used by staff concerned with diesel maintenance; Laira, part of shed yard remains adjoining the main line but no buildings; Plymouth Docks is used as a workshop by Associated British Ports (Fig. P82); St. Blazey's unique half-roundhouse, now a listed building, is intact but out of use.

Wolverhampton Division: Tyseley coal stage, turntable and offices are used by Birmingham Railway Museum; Ludlow is a garage; Trawsfynydd houses a lorry in an Agricultural Merchant's depot.It was a lean-to adjoining the goods shed; Chester had two sheds of which the original G.W. shed is intact but unused while the former L.&N.W.R. shed which passed to G.W. ownership is used for DMU servicing.

Worcester Division: Worcester, as at Laira, part of the shed yard remains but no buildings; Gloucester offices and wheel-drop shed are used by train crews and shed staff; Cheltenham (Malvern Road) is in industrial use; Ross-on-Wye has been converted into a shop; Cleobury Mortimer is intact but out of use.

Newport Division: Aberbeeg is in industrial use; Cardiff Canton has mainly been rebuilt as a diesel depot but the straight-road shed has been retained for diesel servicing; Branches Fork is in industrial use.

Neath Division: Neath N.&B. shed is in industrial use; Danygraig is a rail-connected warehouse; Landore shed has been completely rebuilt as a diesel depot; the wall against the rock face and part of one return end survive at Milford Haven but the site is derelict.

Cardiff Valleys: Radyr is rail-connected and in industrial use; Barry is used for wagon maintenance and repair; Merthyr is a warehouse; Abercynon is in industrial use.

Central Wales Division: Oswestry is in industrial use; Machynlleth is in use for diesel stabling and has been used in recent years for servicing steam locomotives operating special trains; Aberystwyth, the main line shed has been converted to 1' 11½" gauge to house Nos. 7, 8 and 9 of the Vale of Rheidol Light Railway (a private company); Maespoeth (Corris Railway) is used by a railway preservation society; Moat Lane and Pwllheli are in industrial use.

M32 — Column 2, line 2: At some of the former steam sheds converted to diesel depots the fuel storage tanks have been sited in redundant turntable pits, which create the "bunds" to contain oil spillage and leakage required by the Fire Regulations.

M36 — The tables of Locomotive Stock Totals facing this page do not always correspond to the official Annual Totals. No official totals are known for the years 1837-57 and consequently the figures in the tables have had to be compiled from and to agree with the building and withdrawal dates in the relevant Parts. Official records from 1858 to 1870 give totals only with no lists of numbers or names so that it is impossible to define which locomotives were included in the official returns. To complicate matters further, the concept of "Excess Stock" was introduced in 1864 and locomotives in this category were not included in the returns until 1870. The figures for 1858 to 1870 have therefore had to be compiled from building and withdrawal dates, as above, but with the further complication that these dates were confused because some locomotives were added to or taken out of stock in the half year after the official building and withdrawal dates; in addition, the official broad gauge totals for 1859-61 and the narrow gauge total for 1878 are one more than those shown in the table. There were further complications in the changes resulting from the abolition of the broad gauge in 1892. All these discrepancies will be discussed in a reference book on Stock Alterations to be compiled for the R.C.T.S. Library and therefore will not be elaborated here. Finally, the official totals for 1936-47 inclusive show one more locomotive, the Diesel-Electric No. 2 (see Part 11), excluded from the Part 12 table which was confined to steam locomotives.

M38 — Column 1: All six locomotives on loan to the M.&S.W.J.R. went there in

November 1914 and not at dates ranging from July to October as shown. Other loans to this railway are listed on page K233, Part 10. In column 2, loans to L.&N.W.R., No. 142 returned in April 1918 and No. 143 in April 1919. Differences of one month between dates given on page M38 and pages D52 and D66 of Part 4 almost certainly reflect differences between different official records.

M39 — L.&Y.R. No. 1571 was sent to Chester in May 1918 and was returned in July.

M39 — S.E.&C.R. Nos. 58 and 138 were allocated to Bristol, Gloucester, Swindon and Gloucester in that order. No. 350 was at Bristol, Swindon, Malmesbury and Swindon and No. 359 was briefly at Swindon before transfer to Gloucester.

M39 — Column 1, insert below the Tables: A list headed "Foreign engines on the G.W.R. 1920" included ten from three of the other main line railways. Three of these, L.B.&S.C. 0-6-0's, are listed at the foot of page M38 and another five, although listed as if L.&Y.R. locomotives, are the five shown as G.W. (temporary) Nos. 3094-8 on page K273, Part 10; but the list in question shows Nos. 3094-6/8 arriving in 2/20, not 3/20, and all five returning to the L.&Y.R. in 5/20 whereas the Table on page K273, also from official records, show these returned in 1921 and 1922. The other two were L.&S.W.R. 4-4-0's Nos. 739/41, which were on loan to Tyseley shed from 3/20 to 8/20. The list also includes two from "C.R.", both to Westbury in March 1920, of which "No. 6" is shown as moved to Swindon Works in February 1923, while "No. 10" was sent "to Government" in March 1922. "C.R." could either be an abbreviation for "Camp Railway" or "Caledonian Railway"; the latter Company did loan some locomotives to the Government for War service. Even more enigmatic is another, separate, note about two locomotives quoted as "Nos. 1 and 2 ex S.S. Erymanthus, to Swindon Works 12/3/17, to Swindon Stock Shed, August 1917".

The foregoing notes relate to formal loans between railway companies but other reported cases of wartime loans are more likely to have been "hirings" to overcome short-term shortages. Examples are No. 2001, reported on the C.M.&D.P.L.R. at various dates between 11/14 and 5/15 and No. 2195, reported working on the War Department's system at Tidworth in 1940.

M40 — Column 1, lines 17 and 24: The route colour and group letter for the L.M.S. 0-6-0's was "Uncoloured — A" and for the L.N.E. 0-6-0's "Yellow — A" respectively.

M41 — The route colour and group letter for the L.N.E.R. "O4" Class 2-8-0's were "Blue — D", in common with their G.W.R. counterparts.

M41 — Column 2, last two entries and M42, column 1, table of S.R. loans: The route colour for all the ex-S.R. locomotives except No. 478 was Red. The S15 Class 4-6-0's were group letter "D", with the N15X and the I3 4-4-2T letter "C". The solitary Class H15 4-6-0 was on loan for only two months and was not classified in this way.

M42 — Column 1, lines 30-34: Delete and replace by: "Nearly one hundred of these locomotives were allocated to, and about equally divided between, the Wolverhampton and Newport Divisions and the remainder spread over the London, Bristol, Worcester and Neath Divisions."

M43 — Column 2: Delete the last entry shown as U.S.A. No. 2453. This was No. 2353 which should be inserted between Nos. 2352 and 2354 with Date to Stock 2/44 and Date returned 9/44.

M43 — Column 2, line 51: It was not possible to find regular work for all the very large number of U.S. Army 2-8-0's and from about July or August 1943, many of them were placed in store, initially in sidings near Newport Ebbw Junction shed. To avoid confusion, since some of these were part of the same number sequences as those on loan to the G.W.R., and worked on trial in the same area, it will be as well to record the situation in detail. In September 1943, U.S. Army engineers started to prepare the stored locomotives for use, whereupon they went on a trial run to Caerphilly or Pontypridd and then on a 300-mile round trip based upon Ebbw Junction shed, all with G.W. crews. After this they were handed back to the U.S. Army who prepared them for storage in three nearby dumps.

The first dump, at Treforest, was set up by about October or November 1943 and contained 119 locomotives in a single line over 1⅓rd miles long! Apart from two which were stored at Ebbw Junction, the balance, of 234 (making a total of 355 U.S.A. 2-8-0's stored in South Wales) was split between Penrhos (formed about January or February 1944), where there probably were 151, with the balance at Cadoxton, formed about March 1944 and probably holding 83. The stored locomotives left for the Continent in late 1944 and in 1945, some of those at Penrhos being moved to Cadoxton to fill gaps caused by despatches.

M43 — Column 2, line 54: From No. 8432 onwards the number was painted on the buffer beam and the route colour Blue with group letter E painted on the cab side; earlier locomotives were gradually brought into line. On passing into L.M.S. stock, four-figure smokebox number plates were fitted to 56 of the class, with numbers painted on the door in four other cases. This procedure was overtaken by the British Railways renumbering scheme in 1948; No. 8428 received a smokebox plate reading M8428, although the cabside numbers had a letter "M" below them, as in the case of the Western Region's temporary "W" suffix. All the others had five-figure plates, completed in September 1950, M8428 becoming 48428 in March 1950.

None of Nos. 8400-79 received G.W.R. Automatic Train Control apparatus when built but 22 were so fitted when they returned to the Western Region in 1954-56 although, curiously, another seven were not equipped. This work was done in former L.M.S. workshops, where modified vacuum ejectors to suit the higher operating vacuum of Western Region stock also were fitted.

M53 — Alterations to list of purchasers: J. Cashmore's Midlands Depot was at Great Bridge, Tipton (delete "and"), and it has been suggested that they did not cut up any locomotives at Risca. It has been suggested that G. Cohen also cut up locomotives at Rotherwas Estate, Hereford. that Tew & Rhoden were not rail-connected and that their only G.W. purchase (No. 4648) may have been cut up by John Cashmore and that R. S. Tyley of Barry may have been subcontractors for Steel Supply Co.

M54 — Figure P81 showing chimneys made from redundant boiler barrels is reproduced by way of an appendix to the Boiler Appendix. Although the *Great Western Railway Magazine* in which the picture appeared described them as "2301" Class boiler barrels, Appendix Groups 18-21 show that they were fitted to several other classes.

M71 — Note two corrections: re page C18, the reference is to column 2, line 53, not column 1. Re page C50, this should read page C80, column 1, line 53.

M71 — Re C34, column 1, line 26: No. 196 *was* rebuilt at Worcester, incorporating a new boiler built at Wolverhampton.

M74 — Re D82, top of column 2: Typical of the G.W.R. tradition that one member of a class would be different from the rest, No. 2239 definitely never acquired cab side windows.

M76 — Re E52, column 2, line 39: No. 2146 with bunker side steps appears in Fig. P66.

M76 — Re E52, column 2: Nos. 2181-90 continued to be shown on Diagram 0-6-0T B52, probably because the increased brake power was not an item shown on Diagrams.

M76 — Re E53, column 2: No. 2123 was fitted with sliding shutters and cab doors at Swindon in 4/49 so this was not a Derby initiative as Nos. 2134/6 were not done until 4/50.

M76 — Re E53: No. 2034 was still at Caerphilly in January 1959 but had gone by March. It was out of use and awaiting scrapping at Blaenavon in July 1963. Re E54, No. 2069: In July 1957 was reported as transferred to Wolverton Carriage Works for shunting. It was broken up at Swindon.

M76 — Re E54: No. 2092 was out of use, awaiting scrapping, at Bargoed in July 1963.

M77 — Re E76, column 1: The extra pair of short handrails has been recorded on no fewer than 143 members of the class.

M80 — Additions to the table of L.T. locomotives: London Transport records show that Nos. L92, L99 and L89 (G.W. Nos. 5786, 7715 and 5775) were condemned in 9/69, 1/70 and 1/70 respectively. In the footnote *, for "Sheffield" read "Chesterfield".

M81/5 - No. 5723, condemned 11/57, was used as a stationary boiler at West London Oil Gas Plant. Returned to Old Oak Common in 6/59, it had reached Swindon by 11/10/59.

M88 — No. 7402, condemned 4/7/62, was reported working a freight train over the Lampeter-Aberayron branch on 13th October 1962.

M89 — Footnotes: Page M89 reports sale of No. 1600 for scrap in January 1963 (the purchaser was I. R. Morcot Ltd. of Llanbradach) but it was reported out of use and awaiting scrapping at Nine Mile Point on 27/7/63. No. 1607 was cut up on site by G. Cohen, Sons & Co. Ltd. in September 1969.

M89 — No. 1509 was *not* sold from running stock in August 1959. It was stored at Swindon until June 1961 when it left Swindon, in company with Nos. 1501/2, for Messrs Bagnall's works at Stafford. Before any significant work had been done, Bagnall's gave up locomotive work and the three were diverted to Messrs Andrew Barclay's works at Kilmarnock, After overhaul they were delivered to Coventry Colliery, No. 1501 on 2/6/62, No. 1502 at a date unknown, while No. 1509 had arrived by October 1962.

M92 — Re G42, column 1: Delete lines 3-6 from "Nos. 2604/8/29 . ." onwards and replace by: "Nos. 2604/28/9 did receive R.O.D. tenders and were correctly shown on Diagram Q in this respect but these were replaced by Dean (not Churchward), 4,000 gallon tenders of Lots A46/51 (Part 12, page M11). No. 2659, wrongly shown on Diagram Q, had a 3,000 gallon tender from the late 1920's until withdrawn. This accounts for all R.O.D. tenders used with the "Aberdare" Class.

M94 — Re First Edition, column 1, line 17: The slotted-frame bogie ran under Nos. 6014/00/05/21, in that order.

M95 — Re First Edition, page H27, reference to Nos. 2909/29, for "2929" read "2927".

M96 — In the fourth entry for H36, "delete lines 25-40" to read "delete lines 36-40".

M98 — Line 15: Fig. P67 shows No. 7006 with the inverted "U" shape pressing housing the anti-carbonising valve.

M99 — Line 34: Fig. P68 shows No. 4090 with the longer-than-standard smokebox.

M101 — Re Second Edition, list of withdrawal dates: Although No. 7007 had been condemned on 18/2/63 it was noted in service at Hereford on 2/3/63.

M101 — Re Second Edition, column 1, foot of column: The slotted-frame bogie ran under Nos. 6014/00/05/21, in that order. The apparatus fitted to No. 6021 between January 1935 and May 1936 is shown on fig. P69.

M102 — Paragraph headed Single Chimneys — Improved Draughting: After "removed" add: "The longer chimney liner resembled a capuchon all round on those locomotives which retained the standard "King" chimney casing."

M102 — Paragraph headed Mechanical Lubricators: Delete the first line and as far as "those" (i.e., the first five words) in line two and substitute: "Commencing with No. 6022 in February 1948, mechanical lubricators for cylinders, valves and regulators were fitted to all locomotives. When their existing 2-row superheater boilers of Appendix 100(a) were replaced by new 4-row superheater boilers of Appendix 100(b) during the period 1951-56, those" . . . (continue rest of paragraph as printed on page M102).

M103 — Re H29, footnote (d): After *The Pirate* add " . . although the official date is 3/07 and the *Railway Magazine* of August 1907 says that this name was carried until the *Albion* nameplates were replaced on reconversion to 4-6-0."

M103 — Re H31: Revise the list of known fitting of speedometers to read Nos. 6959/60/5/7-9/71/80-2/5/6/93/4/7/8, 7905/6/16/8/21/3. No. 5900 was so fitted by the Great Western Society at Didcot in 1976, no doubt in connection with main-line running. Incidentally, the Society fitted a hopper ashpan to No. 6998 in 1981.

M107 — Re Second Edition, page H38, column 2, lines 27-31: Delete "in January 1955; later this was fitted with a narrow copper cap" from lines 6 and 7 of the page M107 item and replace by: "by October 1954 and this chimney was fitted with a narrow copper cap in January 1955".

M112 — First line to read Nos. 5112/3/40/2. Amplify note * at foot of Table to read: No. 4176 was used as a stationary boiler at Tyseley until May 1967.

M116 —Re K27: The frames of Dowlais No. 22 (ex-G.W. No. 700) were in use as a transporter wagon in July 1949.

M116 — Re K54: David Davies' biography "Top Sawyer" says that *Dove* was bought second hand and delivered in 1857.

M118 — Re K175: *Wellington* was scrapped in or by 1940, not 1944. See amplified detail in the "General Notes" to Part 10, earlier.

M118 — Re K222, column 1, lines 43-9: The implication is that Peckett's No. 475 may not have gone to Llanelly at all, yet it was given the name *Jeannie Waddell* which at least suggests an order by Waddell. Peckett's records seem to show "John Waddell & Sons, Llanelly" crossed out and replaced by "Cumberland Iron Mining & Smelting Co., Millom", with a date 21/9/89. Page M118 quotes another company and a date of February 1889 and both these companies were taken over by the Millom and Askam Hematite Iron Co. Ltd., which was registered on 23/9/90. This discrepancy cannot currently be solved and to add to the mystery, Peckett's records show yet another Class X 0-6-0ST (their No. 482) "18/9/89 to John Waddell & Sons, Llanelly", which has been crossed through and replaced by "Gwaun-cae-Gurwen Colliery Co." No. 482 (with no Waddell family name quoted) was definitely at "G.C.G." in November 1902 as Peckett repaired it for them. Thus there were two Peckett 0-6-0ST's at "G.C.G.", both having a tenuous connection with John Waddell & Sons, with the presumption that No. 475 is more likely to have been with them at some time in view of its name, and the lack of a name for No. 482.

M119 — Line 5: A photograph of W.&L. No. 42, formerly N.&B. No. 3, *Miers*, appears as fig. P70.

M119 — Re K243, column 2: Add to line 6 "the former P.T.R. No. 1 was laid aside awaiting scrapping on 1/6/58".

M119 — Additional to the revision of page K255, column 1, line 32, two six-wheel-coupled locomotives corresponding to Glyncorrwg Nos. 2 and 4 were for sale at auction on 15th February 1926. No. 2, Kitson & Co's No. 5358 of 1921 was of course never a G.W. locomotive but it is interesting to note that having been acquired at the auction by C. D. Phillips of Newport, it turned up in the stock of the Cannock & Rugeley Collieries, whose No. 3, *Progress*, had come from Christopher Rowlands, Swansea Harbour (see page K257). No. 2, named *Thomas*, worked at Coppice Colliery, Cannock, where it was sold for scrap in July 1964. No. 4 was at Neath Shed in August 1926; there is no further trace.

M126 — Table of Railcar withdrawal dates: No. 10 was withdrawn in April 1956, having been completely destroyed by fire (reported on 28/3/56). Add to the top of column 2, page L20, First Edition and after line 9, column 1, page L21, Second Edition: Nos. 8 and 17, withdrawn in January 1959, were intact at Swindon on 7/8/60 and at Tyseley on 8/5/60 respectively.

M126 — Re L21 First Edition and L23 Second Edition: No. 15 arrived at Swindon from Dunball Wharf, Bridgwater on a "Crocodile" wagon on 15/2/51 and an order was given to Works for cutting up on 1/3/51.

M126 — Re L23 First Edition and L24 Second Edition: Nos. 15101-6 were all seen with painted numbers at various dates between 1957 and 1961 (No. 15100 had always had painted numbers).

M127 — Revised Table 2: Diagram M was later amended to include No. 43 and Diagram M1 was cancelled. Diagram T showed the vestibule connections and buffers, which had been omitted from Diagram P although relevant at the time of issue.

M135 to
M177 — Amendments to the Indexes are given in a separate chapter, later.

M168 — Facing: The pair of 8' 10" wheels shown in Fig. M116 are mounted outside the entrance to the National Railway Museum, York, near the Car Park.

PART 13 — PRESERVATION AND SUPPLEMENTARY INFORMATION

Page

N3 to
N15 — The Chapter headed PRESERVATION has been much updated by the new account under this heading but the following individual corrections should be noted:

N3 — Column 2, lines 31 & 33: A 74XX could not have been created genuinely from a 64XX as the former had lever reverse and the latter, screw reverse.

N5 — Column 2, line 2, for 1' 11½" read 2' 6".

N5 — Column 2, line 63: No. 5668 was stationed at Dowlais Cae Harris from December 1929 to July 1938. It did not go to Dowlais in 1983 and is at Blaenavon.

N16 to N18 — The chapter headed SERVICES WORKED BY G.W.R. STEAM RAILMOTORS has been extensively revised by the new Chapter included herein and resulting corrections are set out in the new Chapter.

N19 — In the caption to fig. N12, Wangate should read Wantage.

N20 — G.W.R. Registered Nos. 1 and 2 are dated December 1896 and January 1896 respectively in an alternative list. The owner of No. 2 at the time was Partridge, Jones & Co. There seems doubt about the identity of Registered No. 13 of 1898 as the locomotive *Dos* was Hudswell Clarke's No. 350 of 1889 and there may have been confusion with a later *Dos* which was Hunslet Engine Co.'s No. 453 of 1888.

N22 — Owners' names to be corrected or amended as follows: No. 100, Britannic Merthyr Coal Co.; 115-9, Partridge, Jones & John Paton; No. 111 may have belonged to H. Lovatt & Co., who used Nos. 105-9 on the same contract. There may also be confusion over No. 14(2) because Oliver constructed the eastern part of the Swansea District lines but the locomotive was registered to work between Llangyfelach and Grovesend on the Morriston to Pontardulais section of this project, for which the Contractors were Walter Scott & Middleton, the reputed owners of Manning Wardle's No. 1739.

N22 — The location for No. 114 should read "Maesteg Sidings and St. John's Colliery".

N22 — The registration plate No. 92 of 1911 has also been recorded on Guest, Keen & Nettlefold, Dowlais No. 15, formerly G.W. No. 688 (see footnote (t), page N27).

N22/3 - Amend captions to photographs between these pages: Fig. N19, all the locomotives shown were sold for preservation; fig. N21, No. 5164 was taken approaching Sterns.

N23 — The location for both Nos. 163 and 170 should read "Llanbradach Colliery to Pwllypant Quarry Sidings (coal)".

N28 — Line 31 re EV to read: Ebbw Vale Steel, Iron & Coal Co. Ltd.

N28 — Re H18/H19, wooden nameplate on No. 5076: "County" Class No. 1011 also had wooden nameplates but with raised letters and was almost indistinguishable from the original apart from countersinking of the heads of the fixing bolts.

N31 — Re C78: William Dean unsuccessfully offered the two locomotives and rolling stock to the Penrhyn Quarry Railway in November 1883, so they clearly were not "sold immediately" as stated on page C78. In June 1884 tenders were invited for the locomotives and rolling stock "which may be seen at Festiniog Station", returnable to Paddington by 7th July 1884 and again the response must have been unsatisfactory because it was reported that all the stock went to Swindon in August 1884. According to the late A. C. W. Lowe, the locomotives were then, or soon afterwards, sold "to Ruabon Coal & Coke Co., and then named *Nipper* and *Scorcher*." Ruabon Coal & Coke was the direct descendant of the Ruabon Coal Co., which had been set up in 1856 by a group, dominated by G.W. officers and their friends and chaired by Gooch, to acquire the then existing Brandie Colliery near Ruabon. By 1884, however, Ruabon Coal & Coke had moved to their "Ruabon New" alias "Hafod" colliery, near Johnstown, a colliery which survived until 1968. There was neither record nor memory at Hafod of any narrow-gauge surface railway, nor anything to confirm that the coke ovens, all that survived at Brandie Colliery site by 1884, had any narrow gauge activity in the 1880's or later, If the F.&B. locomotives did move to Ruabon, perhaps they were for use as stationary boilers, which hardly would have carried names. Other researchers claim that the names *Nipper* and *Scorcher* were used in pre-G.W.R. F.&B.R. days.

N31 — Re C90: Midland Railway No. 200A was 0-4-2WT not 0-4-0WT at time of hiring.

N31 — Re G17: There are some discrepancies between different official records of Nos. 3265 and 3365. What appears to be the correct version appears earlier under "Part 7 — page G17". Delete lines 51, 52, 55 and 56 on page N31 and "1930" on page N32.

N32 — Re H30, mechanical lubricator on No. 6967: For "May 1962" read "July 1955".

N33 — Re J14, column 2, line 40: Delete lines 11-17 complete; a correct version appears on page P59.

N33 — Re J52: Owing to reversal of captions on figs. N30/31 facing page N42, "(fig. N30)" in line 44 should be "(fig. N31)" and "(fig. N31)" in line 47 should be "(fig. N30)".

N34 — Re K66/77/85: No. 1068 had G.W. numberplates by January 1923 but retained Cambrian Railways livery; No. 1085 had Cambrian numberplates on 21/9/22 but had acquired G.W. plates by 25/10/22, again retaining Cambrian livery; No. 1086 was not seen after August 1922, when it retained its Cambrian No. 47 and livery. It was stopped in January 1923. No. 1198 had G.W. numberplates when seen in service at Aberystwyth on 24/4/23.

N34 — Re K71: A photograph of the Nantmawr accident on 7/8/08 clearly shows 0-6-0 No. 31 as such, despite what was said in the Oswestry Works record book.

N34 — Re K108: Further information is to hand, see entry K108, earlier.

N34 — Re K218: *Fleetwood* was not "hired" in the conventional sense, see K218, earlier.

N34 — Re K221 and N34: The issue is further confused by Andrew Barclay's written records which show 14" x 20" cylinders and 3' 6" wheels, the cylinder size being at variance with the official drawing!

N35: Reference to L10 (both Editions), footnote (c): Car No. 15 arrived on the Nidd Valley Light Railway in 1921 and was used to provide a public passenger service between Pateley Bridge and Lofthouse until 31st December 1929. The Light Railway station was some half a mile from the Pateley Bridge terminus of a North Eastern Railway branch from Harrogate. The statement on page N35 that the driving cab survived on site has not been authenticated. However, the car had been broken up by Messrs Robinson & Birdsell of Leeds, who had adapted the trailing end of the railcar body with the driving end vestibule as a cabin or store at their yard, where it was photographed in September 1961. This little structure survived the railcar itself by some forty years, being finally broken up in the mid 1970's.

N35 — Re pages L22/3: It has been confirmed that P.W.M. 1622 *was* Simplex 7139. This was first sent to the Engineer's Dept. at Severn Tunnel Junction and Simplex No. 7177 to the Engineer's Dept. at Greenford.

N36 — Line 4: 18 (not 24) H.&C. cars were loaned to the L.M.S.

N36 — Re M37, column 2, line 37: Another source says that Midland Railway 4-6-4T No. 2107 ran light from Bristol to Reading in late April 1913 and ran trials on the S.E.&C.R. for two weeks. If this is correct it must have then returned to Bristol, to be despatched to the S.E.&C.R. again on 21st May, as described on page N36. This source gives no indication of any Great Western involvement which presumably was limited to providing facilities.

N36 — Re M38: The subject of the photograph of a G.&S.W. locomotive at work on the G.W.R. has been positively identified as No. 309 by virtue of its domed boiler; No. 300A never was so fitted.

N36 — Re M48: The reversing gear and motion from *Lord of the Isles*, presented to Victoria Road Secondary School in 1906 and moved to Swindon Works in the 1940's is now in Swindon Museum.

N37 — Re M81: "suffix to read HS not CE" refers to No. 7734, *not* 5734.

N37 — Re M85: Replace reference to 7754 by: 7754 was delivered to Windsor Colliery, Abertridwr but had reached Bedwas Colliery by 9/59, was reported at Llanbradach Colliery in 10/61 and by autumn 1967 had reached Elliott Colliery, New Tredegar (all these places are within a few miles of each other). In 1/69 it moved to Blaenserchan Colliery, Talywain, in the Gwent Eastern Valley and went from there to Mountain Ash in 5/70. Still at Work on 12/6/74 (see Fig. N18, Part 13), it was described as "spare to diesel" in 7/74 and 10/75 and was dumped out of use in 4/79. Presented by the National Coal Board to the Welsh Industrial & Maritime Museum, it was placed on permanent loan to the Llangollen Railway, who took delivery on 6/9/80 (see PRESERVATION).

N37 — Re M87 — Footnotes: No. 9480 was at work on the R.&S.B. dismantling contract on 7/7/65.

N38 — Continuation of M87 — Footnotes: No. 9425 was a working pilot on 10/4/65. Two "9400" Class were seen in Swindon Works Yard without middle wheels on 7/10/67.

N38 —Re M89: No. 1607 was still in use at Cynheidre in June 1968.
N38 — Re M109: Figs. P71 and P72 show opposite sides of Nos. 4257 and 7249 respectively with Type DL boilers.
N39 — Lines 7 and 8 were indistinct in some copies of Part 13. They should read:
M112 — line 34, 4142: suffix to read WD not BH.
M126 — should read "lines 32, 34, 35: add suffix SB from the Table on page M53 to the withdrawal dates of Nos. 15102/3/4.
N41 — Against W.D. No. 132, last sentence to read "Believed *later* in Russian Zone".
N42 — Facing: The captions were reversed. That to N30 should read "Beyer-Garratt 4-6-0 + 0-6-4 (1931)" and that to N31 should read "Beyer-Garratt 2-8-0 + 0-8-2 (1931)."

THE "DEAN GOODS" IN WAR SERVICE

Publication of Part 13 has produced yet more information on this subject, enough to justify another chapter. This starts with Nos. 2308 and 2542 which became Ottoman (Aidin) Railway Nos. 110 and 111 respectively as recounted on page D75 in Part 4. The first Turkish national railway company was set up on 1st June 1927, known as TCDD from the initials of the first four words in the Turkish form of the railway's name. This company bought the Ottoman Railway on 1st June 1935, when No. 110 was renumbered TCDD 33041 and survived until about 1955, having received an Ottoman Railway type of cab (Fig. P74). Ottoman No. 111 was withdrawn in September 1929 and would not have been expected to acquire a TCDD number yet a Great Western tender, minus its centre pair of wheels, was photographed at Erzurum Station as late as 26th September 1976 (fig. P75). It bore the number TCDD 33042, which ex-G.W.R. No. 2542 might have been expected to have carried if it had survived until 1935.

Seven "Dean Goods" never returned from the Middle East and in addition to the two mentioned above, another two were in use on the Anatolian Railway in Turkey in the British Zone, which was under British military control until 1923. In the meantime, these two had become Anatolian Railway Nos. 38A and 83A in that railway's stock list dated 31st December 1922 but they had disappeared by the time that the TCDD had bought the Anatolian Railway in 1928. The remaining three disappeared when the R.O.D. pulled out of Turkey but they may have been taken over when the Greek State Railways (SEK) were formed, as the railways of Salonika were absorbed into SEK and may have been numbered in a series commencing with No. 61.

No. 2578, shown in fig. D144 facing page D88 in Part 4, was not referred to in the text. It was fitted with condensing apparatus and with pannier tanks to receive the condensate, to reduce the visibility of the locomotive when within range of enemy guns but seems to have been the only one so altered, although ten, W.D. Nos. 177-80/95-200 were so fitted in the 1939-45 War. The gear was not effective when the locomotives were worked hard but six, if not all, of the ten retained tanks and gear in the summer of 1944. Removal is not so well documented: Nos. 70177/9/98 had lost theirs by late 1945 and No. 70195 which retained them in May 1946 had lost them by July 1947.

Following up a report of a "Dean Goods" towing three other members of the class through Waverley Station, Edinburgh in 1920 reveals that a total of 56 0-6-0's were sent at intervals from November 1919 to November 1920 for overhaul by private firms. These 0-6-0's have all been described in Part 4 and comprised 43 "Dean Goods", No. 369 of the "360" Class (page D56), six of the "Standard Goods" Class (page D57), of which No. 432 was found beyond repair, and six of the "2361" Class (page D78). The records only identify two recipients: Beardmore of Glasgow had nine and the Yorkshire Engine Co. of Sheffield had twelve. The others are shown as consigned to Barrow-in-Furness (26), Erith (6) and Glasgow (3).

Part 4, page D75 records the return of 46 "Dean Goods" from France in 1919 and it seems likely that contracting-out served the dual purpose of lightening Swindon's workload and of assisting private firms in the transition from wartime to peacetime activity. Also, Swindon would have wanted to give special attention to those returned from France and sent approximately the same number away for overhaul elsewhere.

Official information and personal observations which have since come to light have made it possible to amplify and amend the information on pages N40-44 of Part 13. The notes which follow should be read in conjunction with pages N40-4; it would be tedious to list every amendment.

Dealing first with those which went to France, Nos. 153-5/7-60/85-94 left in March 1940, Nos. 161-6 in April and Nos. 149-52 either at the end of March or early April. Most, if not all, of the 79 sent to France were embarked at Harwich and two of the first (101-8) batch are known to have been landed at Brest. These bore painted names *Troy* and (108) either *Casabianca* or *Casablanca*. This information replaces the queried and incomplete dates on pages N42/3 and the following amendments to pages N41-4 should be noted: No. 102 was at Batignolles Shed, Paris on 31/8/45 and the caption to Fig. D148 facing page D89, Part 4 should be altered to suit; Nos. 115/20/4/5/58/64 were all reported in good condition at Lille Fives Depot on 27/6/46 where they were used on carriage and other shunting duties and nicknamed "Les Churchills". The Lille enginemen described them as "small but very strong". On 3/7/49 advice was received that No. 115 had at last been cut up at Cohen's, Canning Town after lying there for seven months; Nos. 112/4/83/4 were at Brest, out of use, on 29/12/47; and the reference to No. 157's return from Minden, Line 2, column 2, page N40 should include Nos. 111 and 191 as recorded in Table 4. No. 178 was reported at Adisham, Kent on 25/9/43 carrying an obviously informal name *Fagan*. Owing to confusion in R.O. 7/46 between W.D. No. 185 and S.N.C.F. No. 030W016 (which had been W.D. 187), it is not certain which of the two was at Auray. Cottbus against No. 188 in Table 4 is not in Austria, but in Germany. The table of tender numbers on page N44 should include No. 115 with tender No. 130 and No. 164 with tender No. 120.

Page N40 mentions renumbering in the 70XXX series in 1944-6, done to avoid confusion with locomotives of any line in the War zones where they might be called upon to work. Those on the continent at the time were in enemy hands but they were not renumbered when the opportunity came at the end of hostilities (35 of them had been renumbered into the S.N.C.F. series) although contemporary *Railway Observers* quoted 70XXX numbers. There is no record of renumbering in this country but the only known exceptions are W.D. No. 176 which was scrapped before the scheme started and No. 180, so described on the SMR *and* as 70180.

Footnote "g" to Table 4 on page N43 applies also, on photographic evidence, to Nos. 102/4/5/16/20/5/7/8/30/1/49/58/60/4. These were doubtless fitted to shield an open firebox from enemy aircraft. W.D. Nos. 94, 169 and 180 were similarly fitted in this country (fig. P77).

The photograph of No. 160, fig. N9 between pages N22 and N23 *was* taken at Minsk and fig. N10 is of D.R. (German Railways) No. 53.7607, but which cannot be related to any W.D. number, although it has wrongly been linked with W.D. 200, which did not go overseas. No. 188 also was photographed at Orscha, 200 miles west of Moscow, in 1943 and again at Vienna-Hutteldorf in 1948. Finally, "Vapeur, Dampf, Stoom" by Max Delie, published in Belgium in 1985 has two pictures taken at Bruxelles Sud in 1942 showing W.D. No. 106. Thus its move to Poland, if indeed there was one, would have been later than 1941.

More detail is available about the work of Nos. 100/67/8/71/2/4 in North Africa and Italy. All six went first to Algiers in February 1943 with Nos. 172/4 being loaned to the U.S. Army in March for a short time and were returned to the British Army at Bone in Algeria. Nos. 100/67/8/71 went to Tunisia, as already recorded, in June 1943, No. 168 leaving for Bone in November 1943 followed by the other three in January/February 1944. In March 1944 they were shipped to Italy with movements within Italy as shown in the table, with Nos. 171 and 172 being sent on to Pesaro and Rimini respectively in October 1944, where they stayed until at least March 1945 when they were joined by No. 100. By this time they were referred to as Nos. 70100/71/72 with a note in *R.O.* June 1945 that Nos. 70168/74 were "in Italy". Presumably No. 70167 was renumbered at about the same time (but see line 15, column 1, page N40). Although stored out of use from March 1947 they were not officially taken out of Italian Railways' stock until 1953.

Despite the efforts of visitors and other researchers, little more has been found about the "Dean Goods" sent to China. There are no reported sightings and although Chinese practice is to make a "log card" for each locomotive, none have been found.On the positive side, Chinese

Railways published very detailed dimensions lists which reveal that the 0-6-0's were Class XK_3 occupying a number series 61-90, which would almost have been filled by the 25 0-6-0's known to have been sent to China. These were the 22 from France and three from Longmoor (No. 178 although shown as "China? 11/47" does not seem to have been sent there). They are most likely to have been used in the largely agricultural areas of central and southern China, which had a considerable number of light rail lines. These lines have since been upgraded to a 20-tonne axle load which could have made the "Dean Goods" redundant. To comply with Chinese Railways' standards they would have been fitted with automatic couplers and probably with cowcatchers if they were used on main line work.

Owing to a change in the strategic situation, no more were sent to France after April 1940. The 79 already there were Nos. 101-55/7-66/81-94; No. 156 had been held back for training purposes and had reached Longmoor by 7/4/40 but was back at Swindon in July. Work continued on Nos. 167-80/95-9, 200; with relatively few military duties to perform, most of those in this country reverted to their G.W. numbers at this time. These were given officially as Nos. 167-76 but Nos. 156/77/8/96/7/9 were also reported in the summer of 1940 and Nos. 93-100, which came along at the end of 1940 are included in reports of reversion.

With the enemy occupying the west coast of France, rail-mounted guns (most, if not all, stored from World War I) were deployed on railways near the coast in Kent, Essex and Lincolnshire between March and October 1940. The "Dean Goods" were gradually drafted in and at mid-1940, Nos. 169-72/9/80/95-7 were in Kent, Nos. 173-6 in Essex and Nos. 198/9 in Lincolnshire. No. 167 was at Reading in August, No. 168 was not recorded, Nos. 177 and 200 were at Longmoor and No. 178 on the Melbourne Military Railway. When Nos. 93-100 became available at the end of 1940, Nos. 94/5,100 went to Kent and Nos. 93/6-9 to Essex. By 17/11/1940 No. 156 had reached Canterbury, running as G.W. No. 2529. From a report covering the autumn/winter of 1940/1, Nos. 98/9 had moved from Essex to Kent and Nos. 167/8 had turned up, to be joined by Nos. 173/4 from Essex in 1941. Nos. 93, 175/7/8/98-200 arrived in Kent about 1943/4 and were more widely deployed, some being seen as far west as Southampton Docks. Thus all but Nos. 96/7 (and No. 176, prematurely scrapped, see later) worked in Kent during the War period. An example of their work is the Elham Valley line, a since-closed Southern Railway cross-country route between Canterbury and Dover. In 1943, Nos. 95, 175/80/99 were shedded at Canterbury West, servicing one of the guns stationed near Bishopsbourne, hauling the once-daily goods train and working evening "off-duty" trains to Canterbury West station. Of these, No. 180 was still there at the end of November 1944. The two "Colonel Stephens" light railways in Kent also were pressed into service, with the main activity on the East Kent Railway where Nos. 170/1, running as G.W. Nos. 2536/45 were seen in 11/40 while in 1944, Nos. 93/5, 177/97 were recorded at Eythorne. By contrast, the only "Dean" noted on the Kent & East Sussex was No. 197 on 3/8/43 and which had moved to the E.K.R. (above) by 20/7/44. On 2/9/44, Nos. 93/5 were both at Redhill, probably en route for Longmoor. The foregoing are quoted as some of the few precisely-dated movements of the War years. Actually, dispersal of the "Dean Goods" from the south and east coasts to various depots and dock installations had started in the spring of 1941. Brief notes on these locations with the duties performed are appended where known (the information is necessarily incomplete because of wartime secrecy and censorship):

Longmoor. The Longmoor Military Railway, from a junction at Bordon (on the Bentley-Bordon branch of the L.&S.W.R.) to Longmoor Camp, a distance of about five miles, was built between 1905 and 1908/9. A further three miles, from Longmoor to Liss (on the L.&S.W.R. Waterloo-Portsmouth line) was built between 1930 and 1933, although a connection with the S.R. at Liss was not made until 1942. In 1906 the title Woolmer Instructional Military Railway was adopted to stress the training aspect although this reverted to the original title in 1935. Ten "Dean Goods" were stationed here at various times between 1940 and 1947 (see Table 4, Part 13) but it can be mentioned in passing that four locomotives with G.W. associations were used on the railway at earlier dates, as follows: 0-4-4T No. 34 was acquired from a dealer who bought it from the G.W.R. in 1908 and it lasted until 1921; Rhymney 0-6-0ST No. 026 as W.D. No. 101 was here in the early 1920's (see page K108 as amended in this Part); and ex-Taff Vale 0-6-2T, G.W. Nos. 450 and 579 were bought in 12/27 (page K180) and 3/27 (page K175) respectively. No. 579 as *Wellington* was withdrawn by 1940 but No. 450, as W.D. No. 205 (later 70205) lasted until sold in December 1947.

Typical of the difficulty of recording is the discrepancy between accounts reporting only No. 70205 of G.W. vintage on 16/5 and 17/6/44 and again in R.O. 9/44, with the dates for Nos. 70178/99 of 4/44 and 5/44 respectively on page N43. Perhaps they only came in for attention after service in Kent and were out of sight in the shops (70199 left again later in 5/44). *Railway Observers* of the time recorded duties and changes, for example: 4/45, Nos. 177, 70179/95/8 working passenger trains; 5/45, No. 177 had become 70177; 6/45, 70096 had arrived by 13/5/45. 70195 on passenger trains; 7/45, 70177 converted to oil-firing; 12/45, 70096 was fitted with brass G.W. "2425" number plates at Longmoor but they had since been removed; 2/46, reporting visit 12/45, extra side tanks had been removed from 70177/9/98; 8/46, 70177/9/98 at running shed 25/5/46, also 70195 retaining side tanks (10/46 said they had been removed by 19/9/46); 12/46, 70177/9/95/8. 70205 withdrawn 10/46; 8/47, 70177 (oil-fired), 70195/8 dumped in sidings, 70179, withdrawn, in locomotive shed. No. 70178 is not mentioned in any of these accounts and its visit (see above), if any, could have been for repair as a further note says that Nos. 70177/9/98 left Bordon on 15/11/47 for shipment from Birkenhead, bound for China. The last positive record of 70178 was at Bicester W.D. depot in 11/45, which it had left by 14/7/46. The later movements on page N43 are best ignored until the matter can be clarified. The "demonstration purposes" mentioned in lines 50/51, column 2, page D76 (Part 4) seem to have been mainly of purposely derailing, then rerailing — with inevitable damage — and although No. 2531 (as W.D. 70195) nominally outlasted the last G.W. "Dean Goods" proper (No. 2538, condemned 5/57) by nearly two years, it was in no way serviceable. (Fig. P78) No. 2516, which is preserved in Swindon Museum, is more appropriately shown at work in Fig. P79.

Melbourne. Early in the War it was realised that Longmoor would not be able to train all the military railwaymen required and with effect from 19/11/39 the army took control of the L.M.S. Melbourne branch for 9¾ miles southward from Chellaston East Junction on the Trent to Repton line. No. 178 went there in 1940; it did not go to Cairnryan Military Railway in 1941 and may have remained until moved to Burton Dassett. Nos. 93/9, 100 and 171 were there as shown,the last two for only a few weeks.

Early in 1941, work began at two emergency ports in Scotland. Rail connection to one at **Faslane** was put in from the West Highland line near Shandon and W.D. No. 94 (as 2399) was there on 3/4/41. No. 94 later turned up on the **Cairnryan Military Railway** (CMR) which served the other emergency port, Cairnryan, near Stranraer; it left there in August 1942 and was seen in an up goods at Oxford on 11/8/42, no doubt on its way to Bicester. No. 172 also went to Shandon in 1941 and was reported at Burton Dassett on 23/7/42. Others noted at Cairnryan, amplyfying or correcting the table, were No. 93 in November 1941 (running as G.W. 2433), which had reached Old Dalby by 5/42; No. 95 was reported there in 3, 10 & 11/41 and 5/42; No. 170 was there in 12/45 and on 6/4/47 was said to have been "rusting for months"; the only report of No. 199 was on 6/4/47, also rusting away and it had disappeared by August 1947 when No. 170 was not even mentioned in the visit report. No. 171 *may* have been there in 8/41 but was described as with Pannier tanks, which 171 did not receive and may have been confused with No. 177, which *was* so fitted and which was there in 10 & 11/41. Delete "CMR/41?" against No. 95 on page N40 as it *was* there, but delete the same note against No. 178, which was *not* at Cairnryan.

The other W.D. installations where "Dean Goods" were used were mainly dispersed ammunition dumps served by one or more spine railways with numerous sidings, so that an explosion in one dump would not spread to others. While the principal object was transport of war stores via the link to the main lines, the personnel had to be catered for in their leisure time and the 0-6-0's, being vacuum-fitted, were in at least two cases used to work evening and weekend leave trains, not only within the camps but along the main line to the nearest large town or city. The depots, in alphabetical order, were:

Bicester Central Ordnance Depot: Construction began in July/August 1940 and the first trains were worked by No. 156

MISCELLANY — 3

Photograph] [British Railways

Official photograph taken in April 1957 showing No. 4090 *Dorchester Castle* **with the longer smokebox described on page M99.** P68

Photograph] [LCGB Ken Nunn Collection

No. 2931 at Cardiff in 1922/3, showing absence of distance pieces between nameplate and splasher. P69

Reproduced by courtesy of the National Railway Museum, York

Former N.&B. 3 *Miers* **at Limerick as Waterford, Limerick & Western Railway No. 42 on 5' 3" gauge (see pages K237 & M119).** P70

LATE BOILER DETAILS

Photograph] [A. Swain

No. 4257 at Swindon 7/2/60 with Code DL boiler having chimney set forward and larger pressing over oil pipes entering smokebox. P71

Photograph] [P. H. Groom

No. 7249 at Severn Tunnel Junction 4/9/64 with Code DL boiler having chimney set forward (see foot of page N38, Part 13). P72

Photograph] [A. Swain

No. 4171 at Swindon 12/8/62 with Code BC[2] boiler having larger pressing over oil pipes entering smokebox. P73

"DEAN GOODS" FROM BOTH WORLD WARS

Photograph] [T. C. D. D. per George Toms

No. 2308 as T.C.D.D. No. 33041, derelict somewhere in Turkey in the 1950s. Its extended cab had been fitted by the Ottoman Railway. P74

Photograph] [J. E. Buckland per George Toms

An obvious ex-"Dean Goods" tender used as a four-wheeled tank at Erzurum Station, Turkey, 26/9/76. Although numbered 33042 official lists do not reveal any connection with a "Dean Goods". P75

Photograph] [R. H. G. Simpson

W.D. No. 70094 *Monty* **with express passenger headlamps on special train from Bicester Central Ordnance Depot, approaching Oxford L.M.S. Station, 24/6/45.** P76

"DEAN GOODS" MISCELLANY

Photograph] [Colling Turner Photographs Ltd.

W.D. No. 70094 ***Monty*** **at Arncott Shed, Bicester Central Ordnance Depot in 1946, fitted with extended cab roof.** P77

Photograph] [K. Davies' Collection

W.D. No. 70195 as used for re-railing practice, at Longmoor 3/9/49. P78

Photograph] [W. A. Camwell

G.W. No. 2516 (now preserved) in its heyday on a Brecon to Moat lane Junction goods at Builth Road Low Level Station, 15/9/49. P79

in December 1941. One of the few which did not move between Depots, No. 156 was still there on 14/7/46 but final disposal is unknown. No. 94 reached Bicester in 8/42, not 11/43 and No. 95's stay was short as it was not there on 2/3/43. On 28/11/43 Nos. 94, 156/73 were said to be used for passenger work and No. 70094 was photographed on 24/6/45 (Fig. P76). By 11/45, No. 70095 had returned and No. 70178 had arrived, bringing the allocation to five. On 14/7/46 only 70156 was left and in January 1950 two tenders, one lettered W.D. 70094, were seen yet 70094 seems to have been cut up at Hookagate, S.&M.R., over 100 miles away! Delete reference to Bicester against No. 98, page N41.

Burton Dassett Depot was sited on the Stratford-upon-Avon & Midland Junction Railway, between Kineton and Fenny Compton (on the G.W.R. Leamington Spa-Banbury line) and on which construction was put in hand in 1940. A list of locomotives compiled by someone who seems to have been stationed there does not always agree with the Part 13 version and distillation of the evidence suggests that Nos. 169/70/2/4/8 came and went as shown on pages N42/3 but that No. 173 was only there for a very short time, if at all. Nos. 172/4 left for Glasgow, en route for North Africa, on 23/1/43. No. 99 arrived 10/1/43 and left for Wolverhampton 16/4/43. Nos. 169/70 "nicknamed" *Gert* and *Daisy* (the latter name has also been credited elsewhere to No. 197) were used to work thrice-weekly "off-duty" trains to Fenny Compton where a connection was given with G.W. trains to Leamington Spa. Contrary to the opinion expressed by the enginemen of Lille (see earlier), these two were said by the driver-contributor to have been "poor steamers".

Long Marston Depot was connected to the G.W. Stratford-upon-Avon to Honeybourne line and was extensive enough to employ eleven 0-6-0ST but the one-month stay of No. 169 in 9-10/43 appears to have been a loan from Burton Dassett, the significance of which has not been recorded. There is an alternative to the page N41 version regarding W.D. No. 99 (running as G.W. No. 2528). It is said to have come to Stafford Road from Long Marston in 4/43 and to have been sent back there in 7/43. As Burton Dassett and Long Marston were not far apart there could have been short-term loans.

Old Dalby Depot was built on a site generally unsuitable for rail transport due to varying levels and activity seems to have been confined to extensive sidings near the station. "Off-duty" trains were run from Old Dalby to both ends of this Nottingham to Melton Mowbray line but published accounts imply that these were worked by L.M.S. locomotives. A tender locomotive would have been unsuitable for shunting, which may explain why only one "Dean Goods", W.D. No. 93 running as G.W. No. 2433 was reported there on 18/5/42 and on the Melbourne line (see above) still as G.W. 2433, in the *Railway Observer* for September 1943.

The Shropshire & Montgomeryshire Light Railway and several other minor railways had been requisitioned by the Government, at the same time as the four main line railways, on the outbreak of World War II. Control of the S.&M. passed to the War Department on 1st June 1941 and the 28 miles of existing track was thoroughly overhauled and extended by another 50 miles of sidings, including a nine-track exchange siding at Hookagate, where the S.&M. paralleled the G.W.& L.M.S. joint line from Shrewsbury. The remaining sidings included some unusual shapes, see Fig. P80, in order to disperse the ammunition stores as widely as possible.

The 1941 allocation was W.D. Nos. 96/7 (running as G.W. Nos. 2425/42) and 200. No. 200 went away in 1942 but the allocation was increased by the arrival of No. 98 (as G.W. No. 2415) and Nos. 175/6, supplemented by G.W. No. 2462 which was on loan from Shrewsbury shed. W.D. No. 176 is officially recorded as "damaged beyond repair" at Ford, S.&M.R., on 20/10/43 but No. 2558's History Sheet *gives an alternative version*, reading: "Involved in mishap 26/7/43 and reduced to scrap value. Cut up at Swindon, July 1944 but boiler repaired for further use". W.D. No. 93 (as G.W. No. 2433) arrived in the latter part of 1943, possibly replacing No. 176 and apart from a visit to Kent in 1944, was always on the S.&M. In 1943 No. 175 was reported at Canterbury West (see earlier) but was back on the S.&M. by 6/46. W.D. Nos 99, 169/70 arrived in 1944 and the allocation all had 70,000 added to their numbers during the year. No. 70096 was reported at Longmoor on 13/5/45 but was back on the S.&M. by 6/46. Indeed, the S.&M. allocation was more constant than at most depots. No. 70170 went to Cairnryan at the end of 1945 and Nos. 70094/5/180/96/7 had arrived by June/July 1946, the allocation then being Nos. 70093-9/169/75/80/96/7. On 21/6/47, No. "180" was thus numbered on the buffer beam. Scrap dates for Nos. 94-6/8, 175/97 are given in Part 13 and Nos. 93/7/9 had disappeared by 5/48. Nos. 169/80/96 were stored out of use at Kinnerley on 18/9/49 but cutting up has not been recorded.

Finally, No. 70200, stored at Ramsgate until 8/47, was seen hauled "dead" in a north-bound goods at Wrotham, Kent on 20/2/48, in a dilapidated condition.

Names: Between the pressures of active service and wartime secrecy it is not surprising that there are versions of names differing from those given on page N40. One such version is that Nos. 167/8/71/4 were respectively named *Margaret, Rosemary, Voiara* and *Jean Ann* when stationed in North Africa. Nos. 169/70/8/97 (which remained in this country along with the other named examples Nos. 94, 156 and 200) were respectively reported with names *Gert, Daisy, Fagan* and *Daisy* (again!), while one account of Bicester depot says that *Flying Fortress* and *Alexander* were on two different (unspecified) locomotives. Late pictures of Nos. 94 and 169 do not show names and they may have been short-lived but W.D. No. 108 which arrived in France named *Casablanca* or *Casabianca* came back for scrapping still bearing the name!

Final Disposal: Twenty out of the 79 sent to France disappeared without trace but, surprisingly, there is no record of what happened to nine of the 29 which remained in this country in 1940 although obviously broken up as no trace has been found. The known details are:

Sent from France to China	22
Returned to Great Britain for scrapping	29
Known to have been cut up on Continent (Nos. 122/49/66/82/8)	5
Known to have passed behind enemy lines (Nos. 106/32/60)	3
No record; some were run into the docks at Cherbourg and Le Havre by Army drivers, in June 1940, to prevent them falling into enemy hands and one must have become D.R. No. 53. 7607 (see page M74, Part 12)	20
Total sent to France	79
Sent from Great Britain to North Africa, scrapped in Italy (Nos. 100/67/8/71/2/4)	6
Sent to China, 1947 (Nos. 70177/9/98)	3
Scrapped after accident (No. 176)	1
" at Longmoor (No. 70195)	1
" on S.&M.R. (Nos. 70093-9/175/97)	9
Disposal unknown (Nos. 70156/69*/70/3/8/80*/96*/9/200) * Possibly scrapped on S.&M.R.	9
Total in W.D. service	108

THE ORIGINS OF LOCOMOTIVE NAMES

General Principles

Until the advent of the steam locomotive almost the only inanimate objects needing to be identified by name were ships and stagecoaches and, apart from human beings and domestic pets, the only flesh and blood subjects to be named were racehorses. The totals in each case were relatively small and the sources of appropriate names did not dry up. A fairly small number of locomotives drew their names from all three sources, even up to the early 1900's, as described in more detail later.

The object of this Chapter and of Table 3 is to identify the sources of the names used and not to give dictionary definitions, nor the exact location of well-known places, rivers and the like although brief descriptions have been given of most birds and flowers to avoid just saying "bird" or "flower". Where they have been identified, short biographical details have been given of people whose names were used and where they lived in a building which also was the source of a locomotive name, these items have been cross-referenced in the Table. Caution has been exercised in claiming that some names came from racehorses, ships and stagecoaches as *Courier* proves to have been both a ship and a stagecoach, while *Tartar* could have been named after the winner of the St.

LAYOUT OF A TYPICAL WARTIME STORES DEPOT

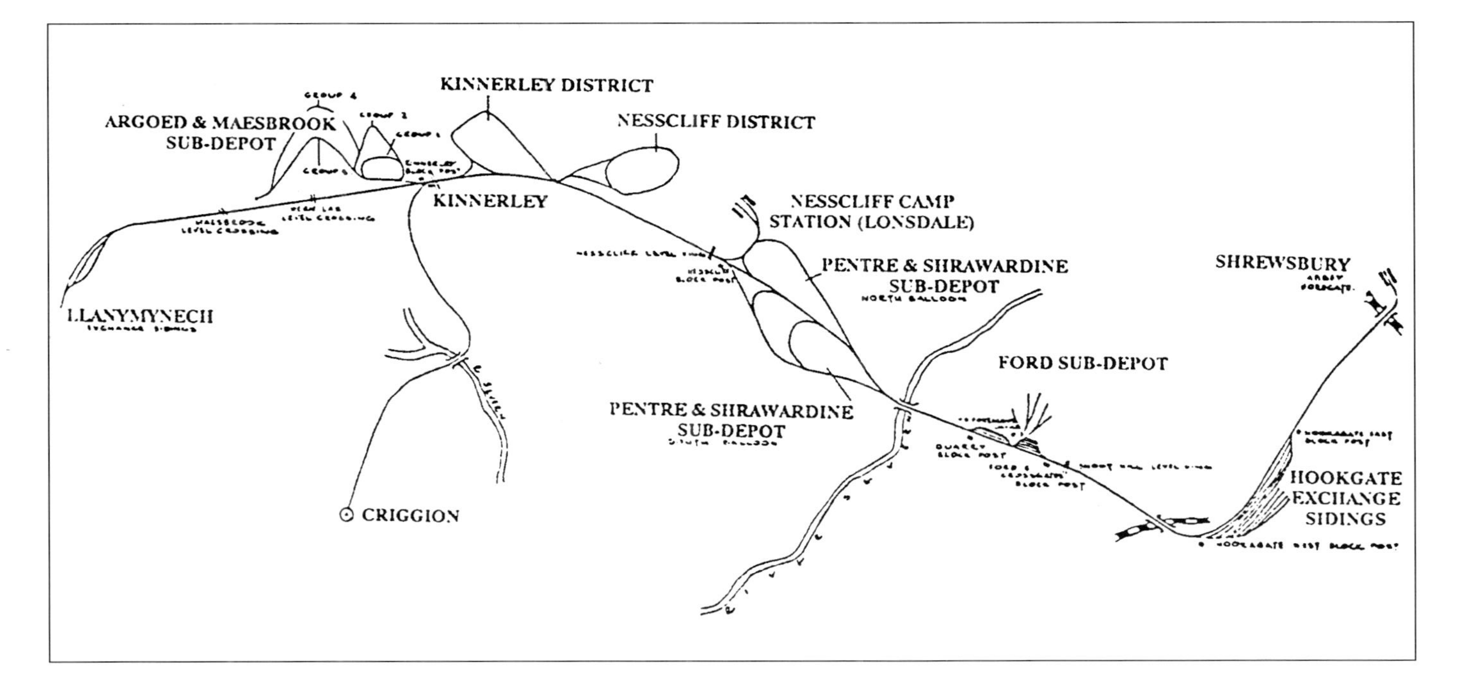

Diagram of the Shropshire & Montgomeryshire Railway at its optimum in the latter part of World War II. Numerous sidings served 205 ammunition store buildings, the system being operated by No. 1 Group of the Royal Engineers. The map and caption have been reproduced from "Branch Line to Shrewsbury — The S & M" by permission of the Middleton Press.

RECYCLING IS NOT A NEW IDEA

Photograph] [G.W.R. Museum, Swindon

No fewer than 48 boiler barrels, stated in the *Great Western Railway Magazine* **for June 1939 to be from "Dean Goods" 0-6-0's, were used to make eight chimneys for the central boiler station at Swindon Works.** P81

Photograph] [David McGuire

Plymouth Docks Locomotive Shed in August 1989, in use as a workshop for Associated British Ports. P82

Leger in 1792 or a ship built in 1840.

Racehorses are the epitome of speed and well known performers on the Turf were an obvious source of evocative names. Some were owned by Directors or Royalty and it was particularly apt that No. 3018, originally named *Racer* should more precisely have become *Glenside* in 1911 when that horse won the Grand National. In the absence of official records it is not possible to be sure that other names attributed to racehorses in the Table were such, hence the use of "probably" but they have been found in Jockey Club records, either contemporaries of the locomotives named after them, or celebrated enough to be remembered long after their day. The railway actually sponsored the "Great Western Stakes" at one time, further proof of the Directors' interest in the Sport of Kings.

Brunel was a ship designer as well as a railway engineer and the name *Great Britain* was almost certainly taken from the 827 ton iron ship which was launched in 1846. His nautical interests may have led him to suggest other ship names and while *Dragon* and *Lightning* are doubtful, the other eight of the first ten of the "Iron Duke" Class of 1847/8 had names to be found on contemporary merchant ships.

Nostalgia seems to have entered into the choice of stagecoach names as some of these were applied more than fifty years after the last coach ran. In the realms of public transport, the coaches were the only long-distance scheduled services before the advent of railways,who even copied their practice of "changing horses" en route. The coaching companies were before their time in promoting their services as the names used identified routes rather than the individual coaches which worked them. Where a stagecoach is the fairly obvious source of a name it is shown as such in the Table. In addition, there is a distinct possibility that *Albion* was the London-Liverpool service, *Blenheim* (London-Oxford), *Courier* and *Greyhound* (London-Birmingham) and *Salopian* (London-Shrewsbury).

It will be wise at this stage to explain that interpretation of the choice of names is not a clear-cut operation. In some cases there are two, or more, potential solutions and where these cannot be solved by context, date or other evidence, the alternatives have been tabled. Names have always been examined in context but there are cases where they do not fit into any pattern. There are complications too where Welsh names have been Anglicised. For example, *Powis Castle* has to be described as situated in the County of Powys, the name of this ancient Welsh kingdom having been revived in 1974, but the Earls of Powis had Anglicised their name and that of their residence.

If the steam locomotive had been designed in the last ten years it would no doubt have been described as "charismatic" and certainly many of the early names were evocative of power, speed, strength, size and noise, including the perhaps inevitable use of superlatives like *Thunderbolt* and *Lightning.* Other names evoked the fire in the firebox and the smoke coming out of the chimney; and on the same principle, very small locomotives (often built for specialised duties) were differentiated from their larger counterparts by names like *Dwarf* and *Tiny.*

To avoid too many headings it is convenient to describe locomotives named to record progress as evocative. The most relevant example of all is *Premier,* the very first Great Western locomotive to be delivered, on 25th November 1837, but *Enterprise, Forerunner, Pioneer* and *Progress* can also be included in this category.

The names given to the first nineteen "miscellaneous" locomotives of 1837-40 were as varied as the locomotives themselves and in some, if not most, cases were chosen by the makers, but the "Star" class of 1837-41 all bore the names of heavenly bodies and, as will be seen from the list, there was some overlapping as four of them were alternative names for *Venus,* which name itself also was carried by two locomotives at the same time. They both lasted until 1870 with no apparent complications although they were known in the books as "Venus No. 1" and "Venus No. 2".

The scope for evocative names was finite and the need for a large number of names in the broad gauge era (and in the corresponding period for the absorbed Companies) resulted in what can only be described as the raiding of classical dictionaries which produced some names that even antiquarian scholars might not recognise. The signs of the Zodiac were an obvious source of inspiration with books on astronomy also invoked in aid. Birds and animals, usually those with a turn of speed, were appropriate, but the choice of half a dozen insects (of which five were noxious and unpopular) looked like scraping the barrel. Ten later members of the same class were more happily named after flowers and shrubs, two of which were wrongly spelt, while there was some confusion about the first four, which seem to have been intended to be named after the national emblems of the four countries. They got *Rose, Thistle* and *Shamrock* right but *Fleur de Lis* (the former Royal Arms of France) seems to have been confused with a very similar device with three flowers similarly arranged, and which is used as a Welsh symbol. It was at one time part of the livery of Cambrian Railways' locomotives, see Fig. K95 in Part 10. The names of battles won, and of those who led us to victory, were appropriate but the choice of ten lethal weapons, hardly relevant to the running of a peaceful railway, seems odd until one looks at the builders, Stothert and Slaughter of Bristol. By 1865 the firm was known as Slaughter, Gruning & Co., and built 20 locomotives in 1865/66 named after contemporary engineers. While this batch was being built, the firm changed its name again to the Avonside Engine Co. but Mr. Slaughter, one of the engineers in question, "autographed" one of those built at the time of the change of name, only to incur the wrath of one of the Great Western directors, who insisted on its being renamed *Avonside.*

The use of people's names had started with the thirteen examples of the "Bogie" class of 1854/55 which commemorated Roman poets and writers. Rather surprisingly this pre-dated the use of "Royal" names (except that *Prince* and *Queen* of 1846/47 may have referred to the Prince Consort and Queen Victoria respectively — her views on the use of his title before hers are not recorded). The first eight of the "Victoria" Class of 1856 bore the names of some of the contemporary rulers but the later ten of 1863/64 took the names of famous engineers, starting, perhaps inevitably, with *Brunel.*

What should have been the first example of an identifiable "Class" name was the "Waverley" Class of 1855, apparently intended to take the names of ten of Sir Walter Scott's novels but the first example, *Lalla Rookh,* was taken from a poem by Thomas More!

A "River" Class of 1857-9 within the "Caliph" Class used the names of eighteen rivers; curiously, seven of these were in North England and Scotland, and three in Ireland while several obvious "Great Western" examples were overlooked.

Finally, the fourteen members of the "Swindon" class, in what looked like a list of the main places served by the G.W.R. managed to include Windsor (at the end of a branch line from Slough) yet omitted Cardiff.

Having found something wrong with naming policy in so many cases on the broad gauge, it is only fair to describe the choice of names for the flagship "Iron Duke" class as inspired poetry. There is no identifiable pattern of names apart from the possible "ship" connection mentioned earlier but they all *look* right.

The late E. L. Ahrons had similar problems and the following is quoted from page 793 of MacDermot's HISTORY OF THE GREAT WESTERN RAILWAY: "Two goods engines received the ridiculous names *Flirt* and *Coquette,* very likely aftermaths of a dance attended by a disappointed member of the staff of the nomenclature office the night before the names were selected". (Actually there was an eighteen month gap between the dates of the two locomotives, so he must, in modern parlance, have been "stood up" twice.) The present writer, infected by Ahrons' frivolous approach, wonders if four other "misfits" of the same period, namely *Florence, Nora Creina, Pearl* and *Ruby* were the objects of his attention. They are all girls' names, quite out of place among the classical, noisy and warlike members of the "Caesar" and "Ariadne" classes. *Nora Creina* has been identified (see Table 3), Florence is an Italian city as well a girl's name, as Pearl and Ruby also are precious stones but none of these names sit happily with the motley crowd just described. (This diversion is intended to show some of the problems which arise in visualising what the people who allotted the names had in mind).

By comparison, the names given to broad gauge locomotives of the absorbed companies are straightforward; the Bristol and Exeter had no named locomotives at all, save for a steam railmotor which was named after the place where it was built. In its last few years the B.&E. had fourteen ex-Great Western 0-6-0's which retained their names, while the South Devon acquired four ex-G.W. six-coupled tanks, three of them retaining their names while the fourth became *Stromboli* instead of *Juno,* thereby still bearing a name which had already been used by the G.W.R.

Indeed, all but five of the South Devon's "main line" locomotives bore names which had already been used on G.W.

locomotives in a similar context and of the five, only *Sedley* has presented any problems. Eight small tank locomotives were named after easily identifiable birds and small animals. *Tiny* was one of those evocative names referred to earlier; the name was apt for a locomotive designed to replace horses on a harbour line! An 0-4-0WT named *Queen* which worked the Torbay and Brixham Branch was no problem and there were plenty of royal princes about to justify naming another small tank *Prince* but the choice of *King* for yet another small tank locomotive could not be justified by historical fact and it must be assumed that it was so named as company for *Queen* and *Prince*!

Another nine assorted broad gauge locomotives came into S.D. stock by transfer on conversion of the railways of their original owners to standard gauge but these are best dealt with under those headings.

The Newquay & Cornwall Junction Railway (later part of the Cornwall Minerals Railway) was worked by contractors, William West & Sons of St. Blazey Foundry. This firm also was closely associated with the local Phoenix mine, which no doubt accounts for one of their locomotive names. Another reputed to have worked on the line was *Roebuck*, said to be the principal promoter of the Cornwall Minerals Railway, presumably in association with Squire J. T. Treffry.

The Vale of Neath was the only other railway (outside those mentioned in the last paragraph above) to have broad gauge locomotives and none of these was named. Neither was any of the 111 "convertible" broad gauge locomotives which the Great Western provided for working broad gauge lines pending conversion to standard gauge.

Standard gauge names in the broad gauge era

The Great Western had an obvious aversion to naming their standard gauge locomotives down to 1892, only four "Royals", three politicians, two stars and either three or four chairmen/directors (there is doubt whether 2-2-2 No. 1130 bore the name *Gooch*), making a total of only twelve or thirteen altogether.

Turning to the naming policies of companies taken over by the G.W.R. down to 1892, identifying the origin of a name in the absence of documentary evidence involves reading the mind of the one who chose the name and context has been found to be the best guide. Where a locomotive was built for any of the Companies who formed the Great Western Railway, the policy behind the name can usually be divined but it has been felt unwise to draw conclusions where other Companies, with their own traditions, are concerned. The prime example is the London & North Western Railway (L.&N.W.R.), which not only sold locomotives to various Companies with names already affixed but also gave its own names to locomotives acquired *from* the Birkenhead Railway in particular. The Manchester, Sheffield & Lincolnshire Railway (M.S.&L.R.) supplied locomotives to a couple of G.W. constituents and the Mersey Railway supplied a number to the Alexandra Docks & Railway Company. Since it is not possible to be certain why these names were chosen they have simply been referred to as "Name chosen by . . ."

Similarly, where locomotives had been acquired from or sold to firms like A. R. Boulton, the locomotive dealer, the sources of the names are unknown; and the most confusing field of all is that of locomotives on short-term hire, or bought from contractors, or sold to industrial concerns where unidentifiable Christian names chosen by the owners were used. Place names sometimes related to the sites of contracts where the locomotive had been used at an earlier date or, in the case of sales out of main line service, to places where they were set to work. We are indebted to the Industrial Locomotive Society and to its members for help in identifying these origins.

Fortunately, identification has been much easier with locomotives owned by Companies which later formed part of the Great Western as they were fairly consistent in giving names of promoters, their families and their residences, and of the places served by the railway in question. Royalty and the occasional politician were honoured while stars, volcanoes, birds and beasts had their place but when they ventured into the realms of mythology, even more obscure reference books than those used by the G.W.'s broad gauge name list compilers gave unidentifiable results.

Some of the absorbed companies had no named locomotives but of those who did, the Shrewsbury & Chester had two "Royals", both specially named for working a royal train in 1852 and a mystery locomotive named *Wrekin*, a name also used by the Shrewsbury & Birmingham for one of its three named locomotives.

The Birkenhead Railway is more difficult. Names of places and rivers, animals and birds, a druid and a monk were no problem. "Evocatives" ranged from two 0-4-0ST predictably named *Cricket* and *Grasshopper* through *Touchstone* (presumably the standard by which other locomotives were to be judged) to *Thunderer* and *Dreadnought*. Names of apparently French origin used for two 2-4-0T of modest size seem more likely to have been those of racehorses. *Voltigeur* won both the Derby and the St. Leger in 1850 (the locomotives were built in 1856); it is not wise to be dogmatic about *Volante*, which won the Oaks in 1792 but one of the Directors may have had recollections of a misspent youth. Determined to be last but not least, four Birkenhead locomotives had names beginning with "Z", all historical or Biblical figures.

The only named locomotives belonging to the Companies which formed the West Midland Railway were *Will Shakspere* (sic) and four fleet-of-foot animals. The Llanelly Railway & Dock Company had a predilection for names of members of the Royal family, five of which were either *Prince* . . or *Princess* . . up to 1859, but from then the names were not prefixed and these five prefixes were removed at about the same time. Having apparently exhausted English Royal names, the Company honoured *Napoleon III* of France in 1868. The remaining names were all of rivers or of places in the area served by the line, with some poetic licence in the case of (Saint) *Teilo*, who gave his name to Llandeilo (usually anglicised as Llandilo).

Owing to problems created by gauge conversion, the Llynfi & Ogmore Railway exchanged its three broad gauge locomotives with the West Cornwall Railway, who sent four standard gauge locomotives with identifiable Cornish names in replacement. There also was a small standard gauge locomotive named *Rescue*, which shunted Porthcawl Harbour, leading to speculation (with no definite grounds) of connection with a disaster at sea off the coast. But even speculation is no help in identifying the broad gauge *Ada*, *Rosa* and *Una*. an example which will illustrate the kind of problem facing the compilers of these notes. Alexander, Henry and James Brogden have been identified as the prime movers in the enterprise and it might be expected that the ladies were members of this family but the family tree contains nobody with any of these names.

The Monmouthshire Railway & Canal Co. did not name its locomotives but did acquire three with names already carried.

Although the locomotive history of the West Cornwall Railway is complicated, most of the names are identifiable except for *Coryndon*, *Ironsides* and *Fox*. There is no trace of an *Ironsides* among a comprehensive extant set of Neath Abbey Ironworks drawings but there were links between these builders and Cornwall and it is just possible that the "Iron" was common to both names. *Fox* was a very small (but long-lived) 0-4-0ST and the name may be evocative of the agility afforded by its 2' 7" wheels!

Locomotives of the Cornwall Minerals Railway (and its later-built Goonbarrow branch) bore straightforward names of directors and places served and it was only when some surplus stock found its way to the Eastern Counties that four of them received names relevant to their new surroundings.

The Carmarthen & Cardigan Railway had a fleeting involvement with locomotives named after birds and volcanoes and the only Company with complications is the Bristol Port & Pier Railway. Its locomotives were not named in B.P.&P. service but reference to both pre- and post-B.P.&P. names has had to be qualified.

Great Western Locomotive Names 1892-1922

From 1892 onwards, most locomotives likely to be found on passenger trains were given names. Like their "Iron Duke" class counterparts on the broad gauge, the "Achilles" class standard gauge 4-2-2's had a very varied and imaginative set of names. Indeed, no fewer than 36 broad gauge names were used again, most of them from the "Iron Dukes". Even the inevitable "royal" names were allocated imaginatively with *Empress of India*, *Royal Sovereign* and a clutch of Dukes and Earls to eke out the "Princess" theme, and with *Windsor Castle* (No. 3080) rounding off the class. Our old friend *Voltigeur* of Birkenhead Railway ancestry (of 1856 vintage) appeared on No. 3059 but the context here suggests the alternative French translation of "Light Infantryman". Although this class was short-lived, no fewer than 17 were renamed; fortunately the reasons for re-naming were fairly obvious.

Changed status of members of the

Royal Family accounted for naming of four more of the "Queen" class 2-2-2-'s in the 1890's and the names given to Nos. 7, 8, 14 and 16 in 1894 honoured the leading lights of the earlier days of the G.W.R.

When the 2-2-2-'s of the 69-76 Class were reconstructed as 2-4-0's in 1895-7 they were named after rivers on the G.W. system with some doubt which River Stour was intended as there are two in the territory. Perhaps the name was chosen hastily as the original intention was to name No. 74 *Thames* but it was found that the nameplate would not fit between the spring hangers. No. 76 *Wye* also has a story to tell as for some years it duplicated the name carried by an ex-Severn & Wye Railway 0-4-0T.

From 1895 onwards there was a tendency to have a theme running through the names given to each class, or parts of a class when the locomotives were built over a period of years. In the case of the 4-4-0's it will be easier to describe them by their original, pre-1913 numbers as renumbering also mixed up some of the classes and put some names out of their original context. For convenience, a summary of the renumbering of the relevant 4-4-0's is: "Duke" Class, post-1913 Nos. 3252-91 were formerly 3252/4-61/5-7/70-2/4-8/81/3-5/7-91, 3313-5/7/9-21/3/6/8/9; "Bulldog" Class Nos. 3300-3440 were formerly 3253/62-4/8/9/73/9/80/2/6, 3312/6/8/22/4/5/7/30-72, 3413-32/43-72, 3701-30; "City" Class Nos. 3700-19 were formerly Nos. 3400-9/33-42; and "Flower" Class Nos. 4100-68 were formerly Nos. 3292-3311/73-81/3-99, 3410-2, 4101-20 (old No. 3382 was scrapped in 1911.) The "Armstrong" Class was assimilated to the "Flower" Class between 1915 and 1923, former Nos. 7, 8, 14 and 16 becoming Nos. 4171/72/70/69 in that order.

The first example of a comprehensive theme was the "Duke" Class of 1895-99, the first forty of which, Nos. 3252-91, all had names of people (real and mythical), places, buildings, rivers and birds directly related to Devon and Cornwall. The later twenty, Nos. 3312-31, maintained this theme up to a point but also embraced Somerset & Dorset, which accounted for seven out of the twenty. The other thirteen names did not fit any particular theme yet more than enough West Country names were used on Nos. 3332-72. The names on Nos. 3332-51 also included other subjects but those on Nos. 3352-72* had a West Country flavour and some of these could have been used earlier to maintain the original theme. (* This assumes that *Sir Stafford* had a West Country connection although Who's Who lists only his then London residence).

Up to about 1899 the Directors chose the names and after that Swindon compiled the lists, subject no doubt to Board approval and this could account for the change of policy. There is evidence that Neville Grenville, a friend of G. J. Churchward, chose the names for Nos. 3332-51.

Reverting to the "Badminton" class Nos. 3292-3311, the names were a mixed batch. They honoured past and present Directors and influential landowners and their residences. *Monarch* might have been a belated compliment to Queen Victoria who celebrated her Diamond Jubilee in 1897 but was more likely a stagecoach, see Table. *Marlborough* may have been a reference to the Duke (or even the college) but the conjunction of *Marlborough* and *Savernake* cannot refer to the branch line, which was opened in 1864! The South Wales Direct Line was in hand at the time; it passed through Badminton and the shortening of the journey to Wales may have inspired the name *Cambria*.

A Swindon ledger of the time has been found, giving three alternative lists of names chosen for Nos. 3373-92, with the major theme of people and places connected with military operations in Africa, almost all of them relating to the Boer War. One of the chosen names was *Conqueror*, to honour one of the British commanders in the Boer War but for some reason there was a last-minute change to (Colonel) *Edgcumbe*, a Director of the G.W.R. An official photograph in workshop grey shows this name (see also page P77 earlier) and the *Conqueror* plates clearly were not carried. *Sir Daniel* was another odd name but in this case the reverse happened and it was changed to the topical South African *Pretoria* after being in service for three months, perhaps because it was realised that the original name duplicated *Gooch* on No. 8 (later No. 4172).

Herschell (see list) had no connection with the Sudan or Boer Wars but the name was said to have been removed early in the 1914 War on the grounds that Herschel (spelt with one "l") was a famous German astronomer!

The whole of the names of Nos. 3393-3412 commemorated a world cruise by the Duke and Duchess of York in 1901. Nineteen of them were named after the ports of call (other than two coaling ports) and *Ophir* was the name of the Orient Line fast passenger ship chartered as a Royal Yacht.

No. 3413 was named to honour the newly-crowned *Edward VII* and Nos. 3414-27 bore the names of Directors although No. 3427 should have become *Fal* in a series of six "River" names. Nos. 3427-32 were to have been *Fal, Plym, Sabrina* (the Roman name for Severn as that name had been used on No. 3328), *Tawe, Usk* and *Yealm*. Under a general re-arrangement, No. 3427 became *Sir Watkin Wynn* and "River" was added to the titles used so that No. 3428 became *River Plym*, No. 3429 became *Penzance* for no obvious reason, Nos. 3430/2 became *River Tawe* and *River Yealm*, while "Usk" was dropped and No. 3431 received the displaced name *River Fal*.

Nos. 3433-42 were named after Cities on the G.W. system. *City of Exeter* was out of alphabetical order on No. 3442, which seems to confirm the view that it was to have been *City of Worcester*. Nos. 3443-52 also took their names from places served by Great Western trains but it is not clear whether the place names were meant to represent the towns served or the names of the stations themselves. The use of the name *Paddington* for No. 3448 supports the latter theory. The intention to name No. 3446 *Liverpool* (which name may not have been carried, see elsewhere) looks like a forlorn attempt to persuade people to use the G.W. route to that city, involving train to Birkenhead, thence by ferry. Wiser counsels evidently prevailed and No. 3446 ran as *Swindon*. The names of Nos. 3443-52 may have been an attempt to promote travel to these places but by 1927 the wheel had turned full circle as names were *removed* over the next 3-4 years because passengers (obviously of the unsophisticated variety) had assumed that the train was bound for the place corresponding to the locomotive's name!

Nos. 3453-72 were named after towns, cities and colonies in or forming part of the British Empire while Nos. 3701-8/20/4 bore names of Directors after running nameless for periods ranging from three to fifteen years. Nos. 3712/29 (as 3422/39) were probably named (in 1913) to promote travel to these seaside resorts, just as *Ophir* was renamed *Killarney* in September 1907 to commemorate the start of the "London and Killarney Day Excursion" trains which started running in that month. The remainder of this batch, Nos. 3709-11/13-9/21-3/25-8/30 never bore names.

The final batch of "Bulldogs" was named after birds, most of them native but with *Flamingo, Pelican* and *Penguin* from overseas.

To complete the outside-framed 4-4-0's, Nos. 4101-20 of 1908 were given names of garden flowers and shrubs, said to have been selected from those grown in Mr. Churchward's own garden, while the hybrid "Earl" Class 4-4-0's were allotted names of past and current Directors and other peers associated with the G.W.R. system. Only thirteen of the twenty names chosen had been used when they were transferred en bloc to the "Castle" Class in 1937. It was said at the time that one of the belted fraternity objected to his name being given to such a small locomotive on humble duties but it is more likely that the decision was made when Stanley Baldwin, a Director from 1908-17 was ennobled at the end of a political career which had included a total of almost nine years as Prime Minister. After running for only a few weeks as *Thornbury Castle*, No. 5063 became *Earl Baldwin* and the names intended for Nos. 3200-19 replaced "Castle" names on Nos. 5043-62.

The first ten of a series of forty outside-cylinder 4-4-0's came out in 1904 with the names of ten English and Welsh Counties through which the G.W.R. passed, but selected on a somewhat random basis and not in geographical order. The next ten, built in 1906, bore the names of Counties in Southern Ireland, no doubt as a promotion for the newly-established Fishguard-Rosslare steamer service. The final twenty picked up the "Great Western County" names which had been omitted in the strict geographical order of the first ten, added the remaining Counties located on the system and then made up the balance from North Wales and the Midlands.

It will be convenient to digress here and to say that the newer "County" Class 4-6-0's of 1945-47 repeated twenty eight of the original English and Welsh County names but replaced Bedford and Flint by Montgomery and Northants. No. 3822 had been *County of Brecon* whereas No. 1007 was *County of Brecknock*. No. 3814 had wrongly been named *County of Cheshire* for the first six months or so and, curiously, the published list of names for Nos. 1000-29 included No. 1011 in the same form but this mistake was corrected before the nameplates were put on.

Before turning to the rest of the 4-6-0's,

reference should be made to the 0-6-4 Crane Tanks built for use at Swindon and Wolverhampton Works, two built in 1901 and the third in 1921. All were given names evocative of strength, reflecting the lifting capacity of the cranes mounted on them.

The locomotives described in this paragraph were renumbered at the end of 1912, the new numbers being shown thus (2900). No. 100 (2900), the first passenger 4-6-0, was named *Dean* (later *William Dean*) in honour of its designer. The second, No. 98 (2998), which was the first true Churchward standard locomotive, ran nameless for four years which is strange because the name *Vanguard* given in 1907 is obviously evocative of a pioneer locomotive design. The third, No. 171 (2971) was appropriately named *Albion* as it was to be tested against the French compound locomotive *La France*. Of the next batch, five numbered 173-7 (2973-7) were built as 4-6-0's and bore the names of Directors. No. 178 (2978), also a 4-6-0, was named *Kirkland* after a racehorse owned by a Director. Nos. 172/9-90 (2972/9-90) came out as 4-4-2's, two named after stagecoaches and three after Directors. These five were renamed after novels by Sir Walter Scott, while the others, except one, had names from the novels, although not when built; and there was the same curious exception, repeating the mistake in the B.G. "Waverley" Class (see earlier) by including *Lalla Rookh*, who was not the subject of a Scott novel. The name of another Scott novel, *The Pirate* may not have been carried by No. 171, see Part 12, page 103 referring to pagc H29.

The remainder of the two-cylinder 4-6-0's settled down to an orderly pattern of what could be called "batch" names, comprising ten Ladies from the pages of history and fiction, twenty lesser Saints (some difficult to identify positively as more than one holy man had the same name during the period from which the names seem to have been derived) and 25 Courts, all but one of which were situate in Great Western territory, some of them the residences of Directors and Officers.

The first four-cylinder locomotive came out as a 4-4-2, was named *North Star* some months later and altered to 4-6-0 in 1909. Ten more 4-6-0's also bore "Star" names, all of which repeated the broad gauge names used in 1837-41 except *Bright Star* and with the same repetition of the "Venus" and "Polaris" themes except that the broad gauge *Load Star* became the 4-6-0 *Lode Star*. The names on Nos. 4011-20 were derived from Orders of Chivalry of which *Knight of the Black Eagle*, a German Order, was removed on the outbreak of war in August 1914. The substituted name *Knight of Liege* was both a Belgian and a French Order but with an acute accent an a grave accent over the "e" respectively (see footnote to Page H11, Part 8, Second Edition).

Nos. 4021-30 took the names of English Kings, starting with the reigning Monarch and evidently intending to work backwards, using each Christian name only once but managed to reverse *King Henry* and *King Richard*. When the "King" Class, Nos. 6000-29 were built (see later), utilising *all* the "King" names back to King Stephen, Nos. 4021-30 were renamed after the Monarchs of nine European countries and of Japan (passing over several Balkan Kings). At first they were (except for No. 4025, which was in Works at the vital time) named *The British Monarch* and so on but after a few months altered nameplates without "The" were fitted. It can only be assumed that as King George V was "the British Monarch", Swindon thought that there could be confusion between Nos. 4021 and 6000. Six of the "Monarchs" lost their nameplates permanently during the early years of the Second World War.

Continuing the "Royal" theme, Nos. 4031-40 took the names of British Queens starting with *Queen Mary* but in nominally working backwards even *Queen Anne*, a Monarch in her own right, was omitted along with several others, perhaps in indecent haste to reach back to *Queen Boadicea* who was here when the Romans arrived. Nos. 4041-5 bore the names of King George V's five sons and Nos. 4046-60 seem to have been named following a general round up of all English princesses extant in 1914.

Having exhausted the "Royal" theme, the final batch of "Stars" took the names of Abbeys situate on the Great Western system although the "Abbeys" at Llanthony and Malvern were technically Priories, having been controlled by a Prior rather than an Abbot. One obvious candidate, *Westminster Abbey*, had been left out of the list and this was rectified a few months later by substituting this for the ruined *Margam Abbey*.

The first of the French De Glehn Compound 4-4-2's was predictably named *La France* but the other two, *President* and *Alliance* commemorated the signing of an Anglo-French Treaty known as "l'Entente

Cordiale'.

The only Great Western 4-6-2 was in many respects an enlarged "Star" Class 4-6-0 but instead of naming it after the largest star in the firmament, Swindon went one better and gave it the name of a five-star constellation.

Absorbed Companies' Locomotive Names, 1892-1913

The Severn & Wye and Severn Bridge Railway's locomotives had a most interesting selection of names; apart from the conventional places on the line, the titles of Crown Officers with duties in the Forest of Dean were used, as were the names of Robin Hood's band of outlaws. These characters are associated with Sherwood Forest in Nottinghamshire and local historians confirm that there is no connection with the Forest of Dean, leaving the reason for the choice unknown.

The Pembroke & Tenby Railway used the names of places on the line, of Directors and their families and of the Chairman's residence in Tunbridge Wells. The North Pembrokeshire & Fishguard Railway started life as the Narberth Road and Maenclochog Railway and local names would be expected to have been given in the vernacular, yet the Preseli Hills became the anglicised *Precelly* and the locomotive which should logically have been *Maenclochog* bore the literal translation of *Ringing Rock*.

The Golden Valley Railway's chequered history was marked by the purchase of second hand L.&N.W.R. locomotives and several hirings from contractors; as mentioned elsewhere it has been felt better not to attempt to identify origins of such names.

The Lambourn Valley Railway's three locomotives took their names from the Saxon version of King Alfred's family names, while both the Manchester & Milford and Liskeard & Looe and Liskeard & Caradon Railways mainly derived their locomotive names from places served by the railways. The L.&L.'s *Lady Margaret* has been identified but not the M.&M.'s *General Wood* and *Lady Elizabeth*. The only clue in the M.&M. case is a letter from Mr. John Barrow to someone unknown, in January 1865 to say ". . . I do not mind if the engine is called *Lady Elizabeth* or *General Wood*", which implies that either was worthy of the honour of having his/her name on a locomotive and could have been man and wife.

Hook Norton seems to have come into G.W.R. stock in settlement of a debt and was named after the site of its operations under its former owners. A few years later it turned up in the stock of the Fishguard & Rosslare Railways and Harbours Company, where it joined company with *Pioneer* (appropriately named for a new venture), *Mermaid* (a name in keeping with harbour construction?), *Wyncliffe* (apparently named after the hotel which later was renamed the Fishguard Bay Hotel) and two named *Nipper* and *Elfin*, presumed evocative of their small size. The only name difficult to identify is *Gallo* which started its working life in Buenos Aires, and the name may have some Iberian connection.

Great Western Locomotive Names, 1923-1950

Up to 1923, names based upon different themes were given to batches within a Class but with the advent of the "Castle" Class all the members of each complete class had names with the same prefix or suffix, namely Castles, Kings, Halls, Granges. Manors and Counties and, except for the "Kings", they all were geographical names although there were a few initial exceptions among the "Castles". These were *Isambard Kingdom Brunel, Sir Daniel Gooch, Viscount Portal, G.J. Churchward* and (fittingly for the last "Castle" of all), *Swindon.*

Five members of the "Star" Class were rebuilt as "Castles" but kept their old names and numbers; the solitary 4-6-2 *The Great Bear* was rebuilt similarly, retaining its unique number 111 but renamed *Viscount Churchill*; and many years later ten of the "Abbeys" were rebuilt as "Castles" but this time they were renumbered in the "Castle" sequence though retaining the "Abbey" names.

To further complicate the matter, one of the "Star" to "Castle" rebuilds was renamed *Lloyd's* and renumbered A1, later 100A1, three of the "Castles" as such had their names replaced by those of Regiments in the British Army (although only one commemorated a particular campaign) and Nos. 4082 *Windsor Castle* and 7013 *Bristol Castle* exchanged identities in February 1952 as No. 4082 was not available to work King George VI's funeral train.

The renaming of Nos. 5043-63 as "Earls" has been mentioned already and the names of Nos. 5071-82 were changed to those of aircraft involved in the Second World War. Two others were renamed to honour General Managers and No. 7005 became *Sir Edward Elgar*, a name previously borne by No. 3414 and much later in British Rail days by a Class 50 diesel locomotive — quite an achievement for one whose only connection with the G.W.R. was that he lived near Malvern. At the suggestion of the writer of these notes, No. 7007, the last express passenger locomotive to be built by the G.W.R. was renamed *Great Western*, repeating the name of the first express passenger locomotive built entirely at Swindon.

All these changes resulted in a pile of redundant "Castle" nameplates which with Swindon's usual penchant for economy would have been stored for possible re-use and it seems likely that they were re-used in most cases. Certainly there is an official instruction to transfer the "Castle" plates fitted to Nos. 5071-80 as built to Nos. 5098/9, 7000-7 and to retain those taken off Nos. 5081/2, but it was too late. The backplates had been used for Aircraft names. Before Nos. 7000/3 were built the allotted names were changed to *Viscount Portal* and *Elmley Castle* respectively and an order given to strip the letters *Cranbrook Castle* and *Drysllwyn Castle* respectively and to use the backplates for the new names. New plates thus had to be made for Nos. 7030/18 respectively. The record for name changes is held by *Denbigh Castle* and *Ogmore Castle*, whose names were given to four different locomotives in succession while nine names were used three times and eleven were used twice.

When Nos. 5083-92 were converted from the "Abbey" series, the original nameplates were retained with a small plate reading "Castle Class" fixed beneath the main plate. This also applied to subsequent "non-Castle" names and the small plates also were fixed to earlier examples of such names except Nos. 4016/37 and 5043-63.

An official list of suggested "Castle" names dated January 1925 identifies all those located on the G.W. system but also includes several sited in the North East and in Scotland, one of which was *Avondale Castle*. These "foreign" castles were not identified as such and it must be assumed that when names were chosen for the 1948 batch, Nos. 7008-17, the then compiler believed from its name that there was, or had been, a castle in the Avondale district in the Cwmbran/Pontypool area in the County of Gwent, then known as Monmouthshire.

The only complications with the "King" Class were renaming of Nos. 6029 and 6028 respectively on the passing of the reigning monarch; the established pattern of working backwards from the current incumbent could not, of course, be followed.

Although there were more than four times as many stately homes with "Hall" names in Great Western territory, history repeated itself in that selection of North Country names for the "River" Class of 1857-9 (see earlier) was followed by looking as far north as Northumberland for the source of names. This started very early in the class, No. 4907 being named after a building in Skipton, North Yorkshire, whose one-time owner had a tenuous connection with the G.W.R. *Cobham Hall* in Kent had been the home of a G.W.R. DIrector and Deputy Chairman and there were a few others with such links but this does not explain why more than one-third of all the chosen names were from outside Great Western territory.

Also, while most of the Halls were conventional stately homes, some were buildings attached to the older Universities and buildings designed for the holding of functions of which *Albert Hall* was the best-known. *Toynbee Hall*, the name of a charitable foundation, was not even a building. From No. 5966 onwards, Swindon recorded the site of the chosen buildings and had done so sporadically with some of the earlier examples but it has been difficult to give a precise location in other cases where several buildings bore the same name and sometimes they were not very far apart. Fortunately there is strong evidence that the G.W.R. used two books, "County Seats" and "Baronial Halls" as the sources of many of the earlier names which has resolved conflicting details but there remain a number which cannot be pinpointed and the only solution here has been to give all the alternative sites in the Table.

The names by which the buildings are now known sometimes vary from those on the locomotives. Among the reasons are changes made by new owners, localised versions (often reflected in the spelling adopted by the compilers of Ordnance Survey maps) and the Anglicising of Welsh names (which also affected some of the "Granges", see later). There were some errors on the original nameplates and Swindon was so meticulous about correcting these even to the extent of inserting missing apostrophes that it is strange that *Runter Hall* was not put right. As discussed in the Table the only "viable" name is Rumer Hall.

The original "Granges" were farms attached to Monasteries (probably derived from the French "grange" = "barn") and 25 of the 80 members of the class come into this category; several were many miles from the parent Monastery. These were usually at or near the places from which they took their names and which were used in four cases for the "Abbey" series of 4-6-0 names. The "Canons of Bristol" and *Llanthony Abbey* were attached to the Augustinian Order; Abergavenny Priory, Gloucester Abbey and Tewkesbury Abbey were Benedictine; Talley Abbey, which had two granges, was attached to the Premonstratensian Order and the remaining 18 locomotives were named after Granges attached to Abbeys of the Cistercian Order. Three of these were *Margam Abbey, Neath Abbey* and *Tintern Abbey.* Flaxley Abbey, of which very little remains, was near Elton in the Forest of Dean. Grace Dieu Abbey, which has disappeared entirely, was near Monmouth, Gwent. The ruins of Abbey Cwmhir are six miles north of Llandrindod Wells, Powys. Four monastic granges became the sites of country houses, the sources of the names of the other 55. Investigation shows that Swindon did not realise that "Llanvair" is merely an Anglicised form of "Llanfair", the Welsh "f" being pronounced like the English "v" and gave the name of the same site to both Nos. 6825 and 6877.

Likewise the name "Manor" originates in Domesday Book, although probably a reference to the existence of Manorial Rights rather than buildings. There were three G.W. examples, two of which now have buildings, but *Cookham Manor* is only a name. The "Counties", the last class of 4-6-0 to be built, have already been dealt with.

Locomotive Names of the Companies acquired between 1922 and 1940

In common with Great Western practice, many of the early locomotives belonging to Companies which later became part of the G.W.R. bore names but not numbers; indeed, there were cases of names only up to absorption in 1923. This was by no means universal and names were not given to any locomotives of the Barry and Rhymney Railways, while the Midland & South Western Junction had a sole, nebulous connection with a locomotive from a Waterworks contract named *Beacon.*

Names carried by locomotives of the Alexandra (Newport & South Wales) Docks and Railway Company were quite a motley collection. The Company's first seven locomotives were bought second hand from the L.&N.W.R., who appear to have cast the nameplates, five of them bearing names of people associated with the railway; the other two, *Rhondda* of 1877 and *Pontypridd* of 1880, were prophetic as there was no link to Pontypridd until 1886. *Aberdare Valley* of 1880 and *Caerphilly* of 1882 were also served by the same link. The former was the only locomotive built *for* the Company which bore a name. Ten locomotives acquired from the Mersey Railway in 1903-5 all had names with, apparently, local significance and it has been felt unwise to speculate on their origin. The other named A.D.R. locomotives were acquired from contractors with equally unidentifiable names except that *Alexandra* could relate to the Company name (but equally could have been named after Queen Alexandra); and *Alexandria* had previously worked on a harbour contract in Egypt. *Sir George* of South Hetton Colliery had no direct connection with the A.D.R.'s *Sir George Elliott* (see page K10, Part 10 and M173, Part 12) and no attempt has been made to identify it.

The early history of the Brecon & Merthyr Railway is linked with Thomas Savin, who also was connected with several of the companies which later formed the Cambrian Railways and transfers of locomotives between the two systems resulted in some with names having a Cambrian affinity working on the B.&M. Mr. Savin, and later the B.&M., also worked the Hereford, Hay and Brecon Railway and some of the names had Wye Valley connections. Four contractors' locomotives named *Stag, Leon, Fairy* and *Dart* are unidentifiable and the reason for the name *Caerleon* (a place miles from the B.&M.), given to a locomotive built as a twin to *Caerphilly* (which *was* served by the railway) is equally obscure. Names of people connected with the line or with the ironworks it served, rivers along the B.&M. and H.H.&B., classical names and evocative names ranging from *Tiny,* through *Pioneer* to *Hercules* were identifiable but the reason for choosing the "foreign" rivers *Severn* and *Mersey* is obscure and it is unclear whether *Cyfarthfa* related to the Ironworks served by the B.&M. or the nearby Castle built by the Ironworks' owners. Finally, in 1908 the B.&M. acquired a former G.W.R. 4-4-0T which had worked at the Ebbw Vale Steel, Iron & Coal Co's premises as their No. 1, *Dickinson*, presumably named after someone connected with that Company. Since the locomotive is reputed never to have been repainted in B.&M. livery it may have retained this name although an oblique-view photograph does not seem to show one.

Until 1899, the locomotives of the Burry Port & Gwendraeth Valley Railway had names only, from 1899 new locomotives had names and numbers (and the surviving earlier locomotives were numbered) but 1909 saw the last named locomotive and a later 1909 locomotive and all later examples bore numbers only. Apart from an early, unidentifiable *Lizzie* the Fairlie locomotives and *Pioneer* all names were derived from places along the line (and the Gwendraeth Valley itself). The first Fairlie locomotive also was named *Pioneer* when built and was changed to *Mountaineer* (an allusion to the gradients on this mineral line) when bought by the B.P.&G.V. The second Fairlie had first been shipped to Australia for use on a 3' 6" gauge line, later returned to this country, altered to 4' 8½" gauge and supplied to the B.P.&G.V. bearing the name *Victoria*, which could be related to its sojourn in Australia or, of course, named after Queen Victoria. The second *Pioneer* came in 1909, the year in which passenger trains were first operated officially, and the name quite likely was connected with the new venture.

The first locomotives to work upon what became the Cambrian Railways, including those involved in the construction process, were owned by the Contractors, Messrs Davies & Savin. As in most other cases, the reasons for naming are usually unclear and it has been necessary to describe them as "Name given by Contractor" or a similar description in the list. Names of places on or near the Cambrian system were given both by the Contractors and, later on, by the Company and are so described in the list but *Milford* (presumably a reference to Milford Haven) seems a long way away to justify this selection.

The Contractors' locomotive *Llewelyn* (known to have been on the line in 1859) must have taken the name of the last native Prince of Wales, Llywelyn Olaf or Llywelyn ap Gruffydd, killed in battle with the invading English in 1282. Llewelyn is the anglicised form and the correct form was actually used by British Rail nearly 100 years later when it was applied to 1' 11½" gauge No. 8 in 1956.

Enterprise was an appropriate name to be given to the first locomotive delivered to the contracts which eventually produced the Cambrian main line; its informal name *Black Donkey* may refer to its paintwork, or its hauling capacity, or both.

Ruthin had worked on the construction of the Denbigh, Ruthin and Corwen Railway and *Whixall* was evidently a reference to Whixall Moss, a bog crossed by the Cambrian Railway east of Bettisfield and which would have caused problems in construction of the line.

Nantclwyd gained its name from a place on the D.R.&C.. where it had worked in company with *Ruthin*. *Merion* and *Cardigan* were anglicised forms of Meirion and Ceredigion, two sons or grandsons of Cunedda Wledig, who ruled most of Wales after the Roman evacuation in A.D. 410. He left the kingdoms of Meirionydd and Ceredigion to them, which later became known as Counties, with the "English" form of Merioneth and Cardigan used at a later date.

Whixall was later renamed *Green Dragon*, part of a heraldic device adopted by the Earl of Pembroke & Montgomery.

Names were given to locomotives built for the Cambrian Railways up to 1878, most of them identifiable as Royalty, Directors and major Shareholders and their residences, places and mountains on or near the line plus one river (the Rheidol) and a clutch of classical names. The Company was politically neutral in honouring both the Whigs and the Tories of the day, none of the four having any known connection with the railway. *Volunteer* seems to have been a tribute to railwaymen who formed part of one of the Montgomeryshire Regiments, perhaps newly returned from service in the Crimean War. *Hero* was hired to work a test train and had no Cambrian connection.

Locomotives acquired from the Lambourn Valley and Metropolitan Railways did not carry names in Cambrian service. For former L.V.R. names see table. Metropolitan were "classical". *Mawddy* was an obvious name for one of the Mawddwy Railway's locomotives but its former owner had named it *Alyn* when used on the construction of the Potteries, Shrewsbury and North Wales Railway which passed through a district so named. The remaining acquisitions, with the Vale of Rheidol and the Welshpool & Llanfair Railways, present no difficulty.

By comparison, the Cardiff Railway was

uncomplicated. A locomotive which had been known as *La Savoie* under a previous owner had worked on a contract in the Savoy area of France and *The Earl of Dumfries* was a complimentary title.

The Cleobury Mortimer and Ditton Priors Railway also made things easy by naming their two locomotives after places on the line, while *Fleetwood* was a locomotive borrowed from the Contractors who built the railway and who in turn acquired it from another contractor based in Fleetwood, Lancashire.

The Corris Railway's only "named" locomotive was not really named at all. The makers' "Class name", no doubt applied for advertising reasons, had not been painted out when it was sent to Machynlleth.

The Llanelly & Mynydd Mawr Railway was worked by Messrs John Waddell & Sons, Contractors, of Edinburgh, who named many of their locomotives after members of the family. *Ravelston* was named after a Waddell Residence at Tumble and the names *Merkland* and *Tarndune*, both of which have a Scottish flavour, may also have been local family homes but these have not been traced. The theme was fortuitously maintained because *Inveravon* had been the residence of an earlier owner of one of the L.&M.M. locomotives. *Great Mountain* was the name of a colliery also owned by Waddell & Sons, and served by the line while *Victory* is an obvious name for a locomotive supplied in 1920. *Seymour Clarke* was acquired already having been named by another Contractor, see Table.

Neath & Brecon Railway locomotives were named after people and places on the line but alternative names *Bulkeley* and *Buckley* quoted for a locomotive hired at a time of shortage do not help identification. Captain Thomas Bulkeley certainly was a Welsh landowner, but not necessarily related to the hirer, a Mr. Forwood, Sir Edmund Buckley owned the Mawddwy Railway but his contractor was R. S. France and the best verdict is "not proven". By contrast, the name *Progress* given to the first locomotive built under Robert Fairlie's patent is self evident and in view of the mountainous gradients on the N.&B., the name *Mountaineer* given to the other Fairlie locomotive is logical in context.

The Cefn & Pyle Railway, a constituent of the Port Talbot Railway & Docks Company, had three named locomotives. The name *Derby* doubtless had a connection with a former owner, J. T. Firbank, who built the Great Northern Railway's Derby-Eggington Junction line. *Penylan* was named after an area of Cardiff in which another former owner, J. E. Billups, did some work for the Taff Vale Railway and *Bryndu No. 3* relates to its former work at Bryndu Colliery.

None of the P.T.R.&D.Co.'s locomotives bore names and neither did those of the Rhondda & Swansea Bay Railway although one bought in from contractors Lucas & Aird was known informally as *Lucas*.

Messrs Powlesland & Mason, haulage contractors at Swansea Docks, had a locomotive with a very long chimney which the staff informally named *Stretcher* and acquired one already named *Dorothy* from a Bilston steelworks. One which became G.W. No. 795 was sold to a steel and tinplate works at Pontardawe and again, by sheer coincidence, was named *Dorothy*. The ladies so honoured probably were unrelated and it is impossible to identify them.

The South Wales Mineral Railway Company's three named locomotives present no difficulty and the locomotives built for the Swansea Harbour Trust had no names. Contractors working for the Trust had an identifiable *Albert Edward*, and *Harriet*, who perhaps was related to Christopher Rowlands but it is difficult to imagine why what must surely have been a small 0-4-0T was named *Jumbo!*

Fifty of the Taff Vale Railway Company's locomotives acquired or built between 1840 and 1863 were named but the nameplates were removed in 1864-6. In most cases the naming policy is identifiable. Three 4-2-0's bought from the Birmingham and Gloucester Railway were American-built which would account for the name *Columbia*, one was *Moorsom*, named after the B.&G.'s Engineer, while *Gloucester* is obvious. *Neath Abbey* was named after the place where it was built and the four names carried by 0-6-0's built by Messrs Stothert & Slaughter of Bristol had a "Bristol" flavour and clearly were chosen by the makers. *Stuart, Bute* and *Dunraven* appear to honour people who had connections with the railway and three animals selected were all fleet of foot, while the Crimean War was topical when *Alma* and *Inkerman* were named. Ten had a mixture of classical and "Star" names but the preference was for geographical names, including rivers which ran parallel with or crossed the railway. The principal places served by the line were included (with *Newbridge* used for the place more usually known as Pontypridd) but in what today would be called "good public relations", many of the collieries and ironworks were included, some of them at the end of short branch lines.

The Cowbridge Railway, a constituent of the Taff Vale, hired four locomotives from I. W. Boulton and it is best not to speculate on the origin of the names, especially as none of the locomotives came into Taff Vale stock. The only other T.V. locomotive to bear a name (and perhaps not for long) was *Radcliffe*, which strange to relate is reputed to have worked on a contract at Radcliffe-on-Trent and on a Manchester Ship Canal contract, not far from Radcliffe, Lancashire and either could have inspired the name.

The last Company in this alphabetical account was the Weston, Clevedon & Portishead, who contributed a locomotive named *Portishead* to Great Western stock. This, and a similar nameless locomotive, had come from the L.B.&S.C.R., where they carried names of places served by that line. Most of the other locomotives bore place names *Weston*, *Clevedon*, or *Portishead* which were used twice, three times and three times respectively.

"Dean Goods" 0-6-0's in W.D. service in World War 2 received temporary names, about half of which are easily identified (including two wartime radio characters *Gert* and *Daisy*) and the remainder seem to have been the names of wives or girl friends of the locomotive crews. According to the Officer in Charge of the Army workshops in N. Africa, the names had been hand-painted on the splashers while the locomotives were still in this country and were painted out in the case of those which were stationed in N. Africa as they went through the shops. There is a variant "eye-witness" account claiming that some of the names were applied in N. Africa but with the traumatic events of the time, records and recollections could be hazy.

There is photographic evidence that W.D. No. 108 still carried the name *Casablanca* (or *Casabianca*, see Table) on return to this country from France for scrapping in 1949.

Finally, if the G.W.R. had lasted into 1948, *Hercules*, the ninth of this name in the Index, would have been a Great Western locomotive. It came from Ystalyfera Tinplate Works in settlement of a debt and as the transaction was made in G.W. days, it deserves a mention.

Tailpiece: Reference has been made to poetic names earlier in this account but there was unconscious alliteration in the choice of three names, all of 4-6-0's. No. 2929 was *Saint Stephen*, No. 4029 was *King Stephen*, and so was No. 6029.

Names proposed but not used

Part 12, Table B, pages M154/5, lists names which had been selected and announced but (with seven exceptions) had not been put on the intended locomotives. Six locomotives ordered for the Cambrian Railways had already been given Welsh place names when the order was cancelled at the last minute. The six went elsewhere (two to the Taff Vale Railway) but the names were removed before delivery. The seventh was *Persimmon*, which has been included in the main list as there is a possibility that it was carried by No. 98 (later 2998) for a short time.

Since the Part 12 list was compiled, several other proposed names have come to light but there is no certainty that the list is complete and the present authors see no point in pursuing derivations. Most of those listed on pages M154/55 had been used before on broad gauge locomotives and will be found in the list and the others are mainly self-evident. Where they help to explain anomalies in the sequence or derivation of names in the list, proposed names have been referred to earlier in these notes.

Montage of three nameplates at Swindon in 1934, from a photograph by P. W. Robinson.

Only *Hedley*, a broad gauge 2-4-0 built in 1865 and later altered to 2-4-0ST, had beading around the name. The others also were 2-4-0's, but of standard gauge, of the "River" and "Achilles" Classes. (See pages B28, Part 2, D43, Part 4 and G11, Part 7 respectively).

TABLE 3

The following Table is designed to be read in conjunction with Tables A and C of the INDEX OF NAMES on pages M154 to M177 of Part 12 as amended or amplified by pages N39 and N44 of Part 13 and by pages P90 to P103 of this Part.

Where a name has been used once only, e.g. *Abberley Hall*, there is no qualification. Where the name has been used on two or more Great Western locomotives, or on two or more absorbed Companies' locomotives, or by former or later owners, or any combination of these, the name is followed by the number of locomotives concerned, e.g. *Aberaman* (2), *Achilles* (3) and *Denbigh Castle* (4). This applies without further qualification where all names have the same source but where there is doubt, or where there are alternative origins, e.g. *Alexander*, these are shown in the Table in run-on form, or cross-referenced to the text on the preceding pages. To avoid overlong entries in the Table, the body of the text has been used to expand upon the contexts, to examine alternative possibilities, or, in the more obscure cases, to set out speculative theories.

Where, for example, somebody's name appeared on one locomotive and his residence on another, there are cross-references in the Table. These are given in italic type like the name on the left hand side (ignore plurals, prefixes and suffixes which sometimes formed part of the locomotive name). Class names are shown as "Bulldog" etc., between inverted commas. County names are those adopted under the 1974 revision of Local Government boundaries.

The same names or initials as those used in Table A, Part 12, pages M137-151 are repeated to identify absorbed Companies' locomotives, e.g. A.D., M.&M. and T.V.

In order to lighten what inevitably is a very long Table, a small selection of pictures illustrating some of the subjects which were the source of locomotive names are included and are identified by (Fig. P83) etc. The pictures of Great Western Locomotive Engineers reproduced as Figs. M2 and M7 in Part 12 are referred to similarly in the following Table.

Abberley Hall	Ten miles NW of Worcester. Now a school.
Abbot	See *The Abbot.*
Abbotsbury Castle	Fortress two miles NW of Abbotsbury, Dorset.
Abdul Mejdid	Ruler of Turkey, 1839-61.
Aberaman (2)	Mining area one mile S of Aberdare, served by T.V.R.
Aberdare (2)	One of the northern termini of the Taff Vale system.
Aberdare Valley	The A.D.R. extended to Pontypridd to tap coal traffic from this source.
Abergavenny Castle	Still stands in the Gwent market town of this name.
Abergwawr	Abandoned colliery NW of Aberaman, served by the T.V.R. at the time the locomotive was built.
Aberporth Grange	Monastic grange near Aberporth, Dyfed, attached to Talley Abbey.
Aberystwyth (2)	The M.&M. locomotive was named after the railway's northern terminus, the Cambrian Coast holiday resort. The "Bulldog" seems to have been named to promote travel to the resort, see text.
Aberystwyth Castle	Well preserved mediaeval castle in this Dyfed town.
Abney hall	At Cheadle, Cheshire.
Acheron (2)	Mythological River of Sorrow in the Underworld.
Achilles (3)	Mythical Greek hero mentioned in Homer's Iliad.
Actaeon	A hunter who was turned into a stag, in Greek mythology.
Active	Evocative name, see text.
Acton	Name given by the L.&N.W.R.
Acton Burnell Hall	Six miles SE of Shrewsbury, Shropshire.
Acton Hall	Buildings of this name are at Sharpness (Glos.), Church Stretton (Salop) and Kidderminster (Hereford & Worcester).
Ada	Could be a sister to *Rosa* and *Una*, but see text.
Adderley Hall	Stood four miles N of Market Drayton, Shropshire. Demolished 1955.
Adelaide	South Australian City visited on Royal Cruise 1901, see text.
Aden	Coaling port at S end of Red sea, visited as for *Adelaide.*
Aegeon	Alternative name of *Briareus*, a monster with 100 arms.

Aelfred	Saxon spelling of the name of King Alfred the Great.
Aeolus	King of Aeolia and controller of winds in Greek mythology.
Agamemnon	King of Mycenae and Leader of Chieftains in the Trojan Wars.
A. H. Mills	Algernon Mills, G.W.R. Director 1905-22.
Ajax (4)	Warrior son of Telemon in Roman mythology.
Alan-a-Dale	Minstrel in the Robin Hood legends.
Albany	West Australian port (in a series of names with British Empire connections).
Albatross	The name of this very large Southern Hemisphere seabird probably evoked the speed, size and power of the locomotive.
Albert (3)	Prince Consort to *Queen Victoria.*
Albert Brassey	M.P. for Banbury, who was a G.W.R. Director 1899-1918.
Albert Edward (2)	Eldest son of Queen Victoria who became King Edward VII.
Albert Hall	Public building in W London, named after Prince Albert, husband of Queen Victoria.
Albion (2)	An old name for England but may have been taken from London-Liverpool stagecoach.
Albrighton Hall	Mansions of this name stand near Shrewsbury and near Wolverhampton.
Aldborough Hall	At Boroughbridge, fifteen miles NW of *York.*
Aldenham Hall	Ten miles NW of Bridgnorth, Shropshire. Now known as Aldenham Park.
Aldersey Hall	Stood near Handley, Tattenhall, Cheshire until demolished, 1958.
Alexander (2)	The B.G. locomotive was named after Alexander II, Czar of Russia (1855-81) and the "Dean Goods" after Field Marshal Earl Alexander of Tunis, C. in C. of the Allied Armies in Italy, 1944.
Alexander Hubbard	G.W.R. Director from 1878, Deputy Chairman 1891-1907.
Alexandra (5)	The A.D. locomotive was named after the Newport Dock served by the railway; the other four the Danish Princess who later became Queen Alexandra and who married the future Edward VII in 1863.
Alexandria	The locomotive had previously worked on harbour construction in Alexandria, Egypt.
Alfred	See *Prince Alfred.*
Alfred Baldwin	G.W.R. Director 1901-05, Chairman 1905-08. M.P. for Bewdley.
Alice	See *Princess Alice.*
Allersley Hall	Incorrect spelling of *Allesley Hall*, see next entry.
Allesley Hall	Coventry Corporation property, three miles W of the City.
Alliance	Name with Anglo-French historical significance, see text.
Alligator	An appropriate name for an early goods locomotive.
Allt	Welsh for "Hill" with the Brecon Beacons foothills in mind.
Alma (4)	A battle area in the Crimean War, 1853-6.
Alyn	River in Clwyd. Name given by an earlier owner.
Amazon (3)	Evocative name based on female warriors in Greek mythology.
Amman	Dyfed river which rises in the Black Mountains and joins the Loughor at Pantyffynnon, also a junction on the Llanelly Railway.
Amphion	Twin son of Zeus in Greek mythology.
Amyas	Cornish sea-captain Amyas Leigh in Charles Kingsley's "Westward Ho!"
Anemone	Woodland and garden flower which flourishes in shady positions.
Anglesea	Acquired from Anglesey Central Railway (alternative spelling).
Annie	Unidentifiable name given by A. Oliver, contractor.
Antelope (5)	Evocative name of a fleet-of-foot animal.
Anthony Manor	Place Manor at St. Anthony, at the entrance to Falmouth Harbour.
Antiquary	Jonathan Oldbuck, Laird of Monkbarns, in a novel by Sir Walter Scott.
Apollo (3)	Greek and Roman Sun God.

Aquarius	11th Sign of the Zodiac. The Water-bearer.
Arab	Evocative of the speed of this breed of racehorse.
Arborfield Hall	Arborfield Court, five miles SE of Reading, demolished about 1949.
Arbury Hall	Three miles SW of Nuneaton, Warwickshire.
Argo (2)	Ship in which Jason made his quest for the Golden Fleece.
Argus	Enormous monster with eyes all over its body (Greek mythology).
Ariadne	Daughter of Minos, King of Crete (ditto).
Ariel	Name apparently chosen by makers who may have thought that this Shakespearian character was the name of a planet.
Aries	1st Sign of the Zodiac. The Ram.
Arley Hall	Six miles NE of Northwich, Cheshire.
Arlington Court	Splendid Regency house six miles NE of Barnstaple, N Devon.
Arlington Grange	At Curridge, Newbury, Berkshire.
Armorel	Mythical maiden said to have lived on *Lyonesse*, which see.
Armstrong	Probably Joseph Armstrong, Locomotive Superintendent at Swindon, 1864-77 but might be George Armstrong who held a similar post at Wolverhampton, 1864-97 (No. 7 was named 3/94) (See figs. M4 & M5).
Arrow	Evocative of the speed of a locomotive and the directness of the railway.
Arthog Hall	At Arthog, near Barmouth, Gwynedd.
Arthur	Prince Arthur, *Duke of Connaught*. One of Queen Victoria's sons.
Ashburnham	Earl of Ashburnham, Chairman of B.P.&G.V. Railway.
Ashburton	Terminus of a branch from Totnes, commemorated by Dart Valley Railway.
Ashford Hall	Three miles S of Ludlow, Shropshire.
Ashley Grange	Once the home of W. G. Grace, Ashley Down Road, Bristol. Demolished 1936.
Ashtead	Station on the L.B.&S.C. system when the locomotive was built.
Ashton Court	At Long Ashton, S of Bristol.
Ashwicke Hall	In the Whitchurch area of S Bristol, and now a school.
Assagais	Plural of Assegai, a lethal weapon (see text).
Astley Green	Colliery between Manchester and Leigh, S Lancashire.
Astley Hall	At Stourport-on-Severn, once home of *Stanley Baldwin*, which see.
Aston Hall	At one time the Birmingham home of *James Watt*, which see.
Atalanta	Mythical huntress brought up by a bear when abandoned by her parents.
Atbara	Egyptian River, tributary of the Nile and scene of battle in Sudanese War.
Athelhampton Hall	15th Century stately home six miles NE of Dorchester, Dorset.
Athelney Castle	Once stood five miles SSE of Bridgwater, Somerset, where earthworks are known as King Alfred's Fort.
Atlas (4)	The first use, in 1838, was as the name of the works where the locomotive was built. The other three are from the mythological Titan, said to be able to hold up the sky and doubtless evocative of the strength of the locomotive.
Auckland	Main city of New Zealand's N Island, visited on Royal Cruise, 1901, see text.
Auricula	Flower similar to the primula.
Aurora (3)	The broad gauge locomotives were named after the Aurora Borealis; the Cambrian/Metropolitan 4-4-0T after the Roman Goddess of the Dawn.
Australia	One of the main British Empire countries when the locomotive was built.
Avalanche (2)	"A large fast-moving mass" was hardly the right choice for a B.G. "banking" locomotive but more appropriate to No. 3003.
Avalon	Name given in the Arthurian legends to a part of Somerset surrounded by marshes.

Avon (3)	In the G.W.R. context, the river joining the Severn Estuary at Avonmouth must have been intended. For the T.V. name see text.
Avondale Castle	Swindon records this as in the (former) County of Monmouth but no such site or building has been found. It was obviously named (inadvertently, see text) after a ruined castle near Strathaven, Strathclyde, Scotland.
Avonside	Re-naming of *Slaughter* for reason given in text.
Aylburton Grange	Monastic Grange near Dymock, Gloucestershire, attached to Llanthony Abbey (technically Llanthony Priory, see text).
Azalia	Wrong spelling of Azalea, a flowering shrub like the Rhododendron.
Bacchus (2)	Mythical Greek God of wine.
Baden Powell	Boer War leader, defender of Mafeking and a national hero in 1900.
Badminton	Duke of Beaufort's country estate. The South Wales & Bristol Direct Railway was being built over his land when the locomotive was named.
Baggrave Hall	Seven miles ENE of Leicester.
Baglan Hall	Three miles W of Port Talbot, W Glamorgan. Once the home of G.W.R. Director *Evan Llewellyn*, it was demolished in 1958.
Balaklava (2)	S Crimean Port, scene of the Charge of the Light Brigade in 1854.
Baldwin	See *Alfred Baldwin.*
Balfour	Lord Balfour, Chairman of Castner Keller Alkali Co. now I.C.I.
Bampton Grange	At Bampton, Oxfordshire, near Brize Norton Airfield.
Banbury Castle	Once stood in the Oxfordshire town but largely destroyed in 1646.
Banshee	Irish fairy attached to a house and bringing evil omens.
Barbados	British possession in the W. Indies in Victorian days.
Barbury Castle (2)	Iron-age hilltop fort five miles S of Swindon, Wiltshire.
Barcote Manor	Eight miles SW of Oxford.
Barningham Hall	12 miles N of Norwich, Norfolk, Known locally as Banningham Hall.
Barrington	Viscount Barrington, GWR Director, Deputy Chairman and Chairman 1840-1863. M.P. for Berkshire.
Barry Castle	Stood near the S. Glamorgan town of that name.
Barrymore	See *Lord Barrymore.*
Barton Hall	Now a holiday centre near Torquay in S. Devon. Partly demolished.
Bath	The B.G. "Swindon" Class bore names of places served by the G.W.R.
Bath Abbey	The present abbey was founded in 1499 in the City of Bath.
Baydon Manor	Near Ramsbury, Marlborough, Wiltshire.
Beachamwell Hall	Five miles SW of Swaffham, Norfolk.
Beacon (2)	Both locomotives worked in the Brecon Beacons area of Powys.
Beaconsfield (2)	Benjamin Disraeli, Earl of Beaconsfield 1804-81. Prime Minister 1868 and 1874-80.
Bear	Informal name given by the staff at Davyhulme.
Bearley Grange	Three and a half miles NNW of Stratford-upon-Avon, Warwickshire.
Beatrice	One of Queen Victoria's daughters.
Beaufort (3)	The first two refer to the Duke of Beaufort, over whose Badminton Estate the G.W.R. built the South Wales & Bristol Direct Railway (see also *Badminton*); the third honoured a bomber aircraft used by the RAF in the early months of World War II.
Beckford Hall	Jacobean building six miles ENE of Tewkesbury, Gloucestershire.

Bee	An industrious insect but hardly evocative of a steam locomotive!
Beenham Grange	Private residence which was once the World War II H.Q. of the G.W.R. It stands nine miles WSW of Reading, Berkshire.
Begonia	A garden flower with coloured perianths but no petals.
Behemoth (2)	Biblical animal, probably the hippopotamus. Evocative of the size of the locomotive.
Belgian Monarch	Renamed to avoid confusion with "King" Class locomotives, see text.
Bellerophon (2)	Mythical Greek God of War.
Bellona	Mythical Roman Goddess of War.
Belmont Hall	School on the E side of Northwich, Cheshire.
Beningbrough Hall	National Trust property eight miles NW of York.
Ben Jonson	English poet and dramatist, 1573-1637.
Benthall Hall	National Trust property four miles ENE of Much Wenlock, Shropshire.
Bergion	Son of the Priestess Hera in Greek mythology.
Berkeley Castle	A most impressive castle overlooking the River Severn midway between Bristol and Gloucester.
Berrington Hall	National Trust property three miles N of Leominster, Hereford & Worcester.
Berry Pomeroy Castle	Old fortress standing on a wooded cliff at Totnes, South Devon.
Bessborough	Frederick Ponsonby was a G.W.R. Director and Chairman, 1857-59 and later became the 6th Earl of Bessborough, an Irish title.
Bessemer	Sir Henry Bessemer invented a process to turn cast iron into steel.
Betty	Temporary informal name given in wartime, see text.
Beverston Castle (2)	Ruined 13th century castle two miles WNW of Tetbury, Gloucestershire.
Bey	Turkish Governor.
Beyer	Partner with Richard Peacock in the firm of locomotive builders.
Bibby	See *Frank Bibby.*
Bibury Court	Nine miles NE of Cirencester, Gloucestershire.
Bickmarsh Hall	Three miles N of Honeybourne, Hereford & Worcester.
Bingley Hall	Birmingham Community building demolished in 1987.
Binnegar Hall	Country house two miles W of Wareham, Dorset.
Birchwood Grange	The home of a G.W.R. Director, SIr W. J. Thomas, in Birchwood Road, Penylan, Cardiff. Now a student hostel.
Birkenhead (4)	The first three uses, all within eight years, referred to the northern terminus of the Birkenhead Railway; the fourth may have been a reference to Birkenhead as a G.W.R. terminus.
Birmingham	The B.G. locomotive took its name from this important junction.
Birtles Hall	Four miles W of Macclesfield, Cheshire.
Bishop's Castle (2)	Stone-built Motte & Bailey castle first mentioned in 1148 but little now remains in the W Shropshire town which bears its name.
Bithon	Son of Argive, a priestess of Hera, in Greek mythology.
Blackbird	Britain's most numerous bird, noted for its melodious song.
Black Prince	Warrior son of *King Edward III.*
Blackwell Hall	A building of this name in London lasted from 1356 to 1820.
Blackwell Grange	Farm at Shipston-on-Stour, Warwickshire.
Blaendare	Name of colliery near Pontypool, Gwent, where No. 789 worked after sale.
Blaisdon Hall	A Roman Catholic School eleven miles W of Gloucester.
Blakemere Grange	Monastic grange nine miles W of Hereford, attached to Abbeydore.
Blakeney	The new owners of this C.M. locomotive named it after a village eight miles W of Sheringham on the N Norfolk Coast.
Blakesley Hall	At Yardley, Birmingham, now the Birmingham Museum & Art Gallery.

Blanche — Blanche Vere Guest, youngest sister of *Sir Ivor* Guest, which see.

Blasius — Latin form of Blaise, the patron saint of a church near St. Blazey, Cornwall.

Blazer — Evocative name, see text.

Blenheim (2) — The "Badminton" Class locomotive bore the name of an Oxfordshire country estate given to the first Duke of Marlborough by Queen Anne after the battle of Waterloo; the "Castle" honoured a fighter-bomber used in the early years of World War II.

Blenkensop — John Blenkinsop (wrongly spelt on locomotive), a pioneer locomotive builder of Leeds.

Bodicote Grange — Oxford Road, Banbury, Oxfordshire, now demolished.

Bodinnick Hall — Near the mouth of the River Fowey, Cornwall, on the E bank but little now remains.

Bombay — Indian Port and State Capital (British Empire connection).

Bonaventura — Shepherdess from Week St. Mary, North Cornwall, who went to London as a servant girl and married three times. Her third husband became Lord Mayor in 1498. She left money in her will to pay for a College and Chantry Chapel in her home village.

Borth — Between Aberystwyth and Machynlleth, served by Cambrian Railways.

Borwick Hall — Two miles NE of Carnforth, now the training centre of the Lancashire Youth Clubs Association.

Boscawen — Family name of Lord Falmouth.

Bostock Hall — Three miles NW of Middlewich, Cheshire. Now a school.

Bourton Grange — Three miles SW of Much Wenlock, Shropshire.

Bouverie — Name given by the Mersey Railway.

Bowden Hall — Three miles SE of Gloucester and now a hotel.

Boyne — River in SW Ireland, see text.

Bradfield Hall — Six miles W of Reading and a mile SW of the village of Bradfield.

Bradley Manor — Newton Abbot, South Devon. Built in 13th century. National Trust property.

Brasenose — One of the Oxford Colleges and landlord of some G.W.R. land.

Breccles Hall — Ten miles NE of Thetford, Norfolk, known as Breckles Hall locally.

Brecknock — Pre-1974 Welsh County, now part of Powys.

Brecon — Northern terminus of the Neath & Brecon Railway.

Brecon Castle — Well-preserved ruin in the town of Brecon.

Briareus — In Greek mythology, a monster with 100 arms.

Bricklehampton Hall — Regency house on the slopes of the Bredon Hills in Hereford & Worcester. Now a nursing home.

Bride of Lammermoor — Character in a romantic novel by Sir Walter Scott.

Bridgwater Castle (2) — Stood in the W of the Somerset town of that name. Demolished after capture in 1645, leaving only fragments of stonework.

Brigand — The S.W.M. locomotive was of unknown origin but, like the G.W. example, must have been "one of a gang of bandits". (The twin of the G.W. example was aptly named *Corsair*).

Bright Star — A general name for bright celestial objects, e.g. *Venus* and *Jupiter*.

Brindley — James Brindley (1716-72), well-known canal builder.

Brisbane — Capital of Queensland, visited on Royal Cruise 1901, see text.

Bristol (3) — The B.G. locomotive took its name from the original western terminus of the G.W.R. The T.V. example was built in Bristol and was named by the makers.

Bristol Castle — Fortress which stood on the banks of the Avon from 887 and is portrayed as part of the Great Western Coat of Arms.

Britannia — Roman name for Britain, personified by a seated and helmetted lady who became a symbol and was notably found on the reverse of a pre-decimal penny. The A.D. example may have been named by a previous owner.

British Monarch	Renamed to avoid confusion with "King" Class locomotives, see text.
Brocket Hall	Stately home two miles SW of Welwyn, Herts, often used as a conference centre.
Brockington Grange	At Bredenbury, near Bromyard, Hereford & Worcester.
Brockley Hall	In the SW suburbs of Bristol.
Brockton Grange	Swindon records this as near Newport, Salop, and the most likely is at Brymhill, six miles SE of Newport although there is another near Shifnal, eleven miles due south of Newport.
Broneirion	Half a mile west of Llandinam, Powys, residence of David Davies who partnered Thomas Savin in building most of the Cambrian Railways.
Brontes	In Greek mythology, a blacksmith who was one of the *Cyclops.*
Broome Hall	Swindon records show this at Holmwood, near Dorking, Surrey.
Broughton Castle	This fortified manor house three miles SW of Banbury, Oxfordshire, has been the home of the Lords Saye and Sele since 1451.
Broughton Grange	In Windsor Road, Banbury, Oxfordshire.
Broughton Hall	At Skipton, N Yorkshire. Owned by a branch of the family of Lord Vane-Tempest, killed in the Cambrian Railways Abermule disaster in January 1921.
Browsholme Hall	Five miles NW of Clitheroe, Lancashire.
Brunel	First Engineer of the G.W.R. See *Isambard Kingdom Brunel*, and Fig. M2.
Brunlees	Name given by Mersey Railway.
Brutus	Legendary founder of the Roman Empire.
Bryngwyn Hall	One mile ENE of Tremeirchion, near Mold, Clwyd. Now St. Beuno's College.
Bryn-Ivor Hall	Stood at Marshfield, SW of Newport (Gwent). Demolished 1950.
Buckenhill Grange	At Bromyard in Hereford & Worcester.
Bucklebury Grange	Five miles NE of Newbury, Berks.
Buckley	Possible alternative name to *Bulkeley* on the N.&B., see text.
Buffalo (3)	The powerful American bison, evocative of the power of steam.
Builth Castle	At Builth Wells, Powys.
Bulkeley (4)	Captain Thomas *Bulkeley*, Welsh landowner and G.W.R. Director, 1850-82.
Bulldog (2)	The name of the Golden Valley's locomotive was given by the L.&N.W.R.; a class of 156 4-4-0's was known by the name of an early member, so appropriately evocative of power.
Bullfinch	A familiar red-pink bird, well known for its love of fruit buds.
Bulliver	Nickname for a locomotive or train on the Totnes-Ashburton branch.
Bulwell Hall	Public building in Nottingham demolished in 1958.
Burcot	Name given by the Mersey Railway.
Burghclere Grange	Four miles S of Newbury, Berkshire, and now a school.
Burmington Grange	Two miles S of Shipston-on-Stour, Warwickshire.
Burntisland	Waddell & Sons were building the North British Railway's Inverkeithing to Burntisland line when this locomotive was supplied.
Burry Port (2)	Four miles W of Llanelli, Dyfed. Harbour from which coal conveyed by the B.P.&G.V.R. was shipped.
Burton Agnes Hall	Stately home six miles WSW of Bridlington, Humberside.
Burton Hall	Four miles NE of Loughborough, Leicestershire.
Burwarton	Two miles SE of Cleobury Mortimer, Shropshire.
Burwarton Hall	At *Burwarton* (see above), home of Viscount Boyne who had family connections with the C.M.&D.P.L.R.
Bury	Edward Bury (1794-1858) founder of a Liverpool firm of locomotive builders.
Bute	Cardiff West Dock, built by the Marquess of Bute and opened in 1839 provided the Taff Vale Railway with coal exporting facilities.

Butleigh Court Two miles SSE of Glastonbury, Somerset. Home of Neville Grenville, a personal friend of Daniel Gooch and his successors at Swindon until he died in 1936. He had influence in the choice of names, see text.

Butlers Hall At Lechlade, Gloucestershire. Now known as Butlers Court.

Cadbury Castle Although a building so named exists near Yeovil, Swindon records show the origin of the name as an ancient earthwork at Thorverton, eight miles N of Exeter.

Cader Idris Impressive mountain range three miles S of Dolgellau, Gwynedd.

Cadmus Name given by L.&N.W.R.

Caerhays Castle Coastal castle in Veryan Bay, near St. Austell, Cornwall.

Caerleon Roman town two miles NE of Newport, Gwent.

Caermarthen Alternative spelling of Carmarthen, reflecting the Welsh name for the town (Caerfyrddin) which means Merlin's Camp.

Caerphilly (2) Like *Caerleon* (above), Caerphilly was not on the B.&M. system; the A.D. locomotive worked trains between Newport, Caerphilly and Pontypridd.

Caerphilly Castle Splendidly restored 13th century castle in this Welsh town (Fig. P89).

Caersws Site of junction of Van Railway with Cambrian Railways main line.

Caesar Title given to Roman Emperors.

Calceolaria Originating from South America, they have curious pouched flowers.

Calcot Grange Beside the G.W. main line to the west, just beyond Reading. The retirement home of *Sir Felix Pole*, General Manager 1921-9.

Calcutta Capital of Indian state of W. Bengal (British Empire connection).

Caldicot Castle Well preserved castle at Caldicot, Gwent.

Caldicott Castle Mis-spelling of the previous name.

Calendula Presumably Calendula officianalis, a species of *Marigold*.

Caliban Shakespearean character who sits oddly among members of a Class named after large animals and a mythical blacksmith.

Caliph Middle East chief civil and religious ruler deriving his authority as a successor to Mohammed.

Calveley Hall Six miles N of Nantwich, Cheshire.

Camborne (2) West Cornwall Railway station four miles west of Redruth.

Cambria (3) The Roman name for Wales, see also text.

Cambrian (2) Almost certainly a colliery served by the T.V.R.

Cambyses King of Persia, 530-522 B.C.

Camel (2) The S.D. Class bore names of strong animals, men and reptiles; the "Bulldog" was the Cornish river which enters the sea at Padstow.

Camelia Mis-spelling of *Camellia*, which see.

Camellia Flowering shrub introduced to this country in the 18th century.

Camelot Legendary West of England site of King Arthur's Court.

Campanula Species of garden flower, including the bell-flowers.

Campion Hall Permanent residential hall attached to Oxford University.

Canada Locomotive used by the contractor who worked goods traffic on the C.M.&D.P.L.R. until the Company's own locomotives arrived.

Canary Name given by contractor to a locomotive used on the O.W.&W.R.

Cancer 4th Sign of the Zodiac. The Crab.

Capel Dewi Hall Four miles E of Carmarthen, Dyfed.

Capesthorne Hall Five miles W of Macclesfield, Cheshire.

Cape Town Capital of Cape Province in S. Africa. (Visited on Royal Cruise, 1901, see text).

Capricornus 10th Sign of the Zodiac. The Goat.

Caradoc Grange Two miles NE of Church Stretton.

Caradon	Terminus of Liskeard & Caradon Railway.
Cardiff (3)	Important station on G.W.R. system; hub of Taff Vale Railway; and a name given by A.R. Adams of Newport, dealers who rebuilt ex-G.W. No. 933 in 1932, presumably named for publicity purposes.
Cardiff Castle	Magnificent and extensive castle in the City Centre
Cardigan	Anglicised form of Ceredigion, first ruler of the County, see text.
Cardigan Castle	Two towers of this 11th century castle still stand above the River Teifi in this Dyfed town.
Carew Castle	Five miles NW of Tenby, Dyfed.
Carmarthen	Nominal southern terminus of the M.&M. through running powers over the Carmarthen & Cardigan Railway. See also *Caermarthen* above.
Carmarthen Castle	Ruins still stand in this capital town of Dyfed.
Carnation	One of the dianthus family of garden flowers.
Carn Brea (2)	Site of the West Cornwall Railway's workshops.
Carn Brea Castle	Ancient fortress S of the G.W.R main line just W of Redruth, Cornwall.
Casablanca	(or *Casabianca*). Temporary informal wartime name, possibly the place but more likely the name of a popular song, see text.
Castell Deudraeth	Residence of D. Williams, Chairman of Aberystwyth and Welsh Coast Railway.
Castor (3)	Twin son of the King of Sparta in both Greek and Roman mythology.
Cato (2)	Dionysius Cato, 3rd/4th century Roman poet.
Cavendish	Name given by Isaac Watt Boulton, see text.
Caxton Hall	Public Building in Westminster, London.
Caynham Court	Three miles SW of Ludlow.
Cefntilla Court	Ten miles SE of Abergavenny, Gwent.
Centaur	One of a number of creatures of this name with body and legs of a horse together with torso, head and arms of a man. Greek mythology.
Cerberus (3)	Monstrous watchdog at the entrance to the Underworld. Greek mythology.
Ceres (2)	Roman Goddess of Corn.
C. G. Mott	See *Charles Grey Mott.*
Chaffinch	Common British bird surpassed in numbers only by the *Blackbird*, Robin and Wren.
Champion	Name given to a B.G. goods locomotive for no obvious reason.
Chancellor	Sir Stafford Northcote, Chancellor of the Exchequer, visited Stafford Road Works when the locomotive was being renewed.
Chandos	Name given by the L.&N.W.R.
Chanter	Contractor who worked the Hayle Railway in West Cornwall.
Charfield Hall	Nine miles N of Chipping Sodbury, Avon.
Charles Grey Mott	G.W.R. Director 1868-1905. One of an engineering family.
Charles J. Hambro	G.W.R. Director 1930 and Chairman 1940. Resigned in 1945.
Charles Mortimer	G.W.R. Director 1890-1923 and descendant of the Mortimers, Earls of March, who led their forces against *King Henry IV* in Welsh Border Wars of the 14th/15th centuries. Owners of *Wigmore Castle* at that time.
Charles Russell	G.W.R. Chairman 1839-1855. M.P. for Reading 1830-1837, and 1841-1847. Lived at *Swallowfield Park*, near Reading.
Charles Saunders (2)	Secretary of G.W.R. London Committee 1833-1840, Secretary and General Superintendent of the Line 1840-1863.
Charon	Ferryman on the River Styx between Earth and the Underworld. Greek mythology.
Cheesewring	A village now known as Minions, on the former Liskeard & Caradon Railway.
Chepstow Castle (2)	Ruined castle in the Gwent town of this name.
Cherwell Hall	Part of St. Hilda's College in SW Oxford.
Chesford Grange	One mile S of Kenilworth, Warwickshire. Now a hotel.

Chester (3)	County town of Cheshire; the southern terminus of the Birkenhead Railway; and name given by the Commissioners of Chester General Station to the locomotive which became G.W. No. 342 in 1865.
Chester Castle	Built in 1069 and now the museum of the Cheshire Regiment.
Chicheley Hall	Three miles NE of Newport Pagnell, Buckinghamshire.
Childrey Manor	Three miles W of Wantage, Oxfordshire.
Chirk Castle	14th century castle one mile W of Chirk Station.
Chough	A crow with red bill and legs, named after its call. Once common on the cliffs and around the mines of Cornwall, it is the Cornish emblem.
Chronos	Titan from Greek mythology, and son of Heaven and Earth.
Churchill (2)	See *Viscount Churchill. Walton Park* was renamed by a later owner, see page P75.
Cicero	Marcus Cicero, 106-43 B.C. Roman Orator and Philosopher.
Cineraria	Garden flower which originated from S. Africa.
City of . . .	Ten cities on the G.W.R. system were chosen as names for Nos. 3433-42 (later 3710-9) and one, *City of Birmingham*, was one of the informal names given to a "Dean Goods" 0-6-0 in World War II.
Claughton Hall	Stands neart Garstang, NW of Preston, Lancashire.
Cleeve Abbey	12th century Abbey near Washford, Somerset, of which the priory remains.
Cleeve Grange	At Bishops Cleeve, Cheltenham, Gloucestershire. Bought with a view to redevelopment in 1989.
Cleobury	Cleobury Mortimer, Shropshire, S terminus of the C.M.&D.P.
Clevedon (3)	Mid-point of the W.C.&P.
Clevedon Court	Ten miles SW of Bristol, where Thackeray wrote "Vanity Fair".
Clifford Castle (3)	11th century castle overlooking the River Wye near Hay-on-Wye, Hereford & Worcester.
Clifton	The makers named this T.V. locomotive after a Bristol suburb.
Clifton Hall	Not identified in G.W. records. There is one near Worcester, one at Nottingham and one at Clifton Campville, Staffordshire.
Clun Castle	On the Welsh border N of Knighton, Shropshire.
Clyde	River in W Scotland on which Glasgow stands, see text.
Clyffe Hall	One mile SW of Market Lavington and five miles S of Devizes.
Coalbrookvale	Name of colliery at Nantyglo, Gwent, operated by purchasers of this locomotive.
Cobham	Viscount Cobham, formerly Lord Lyttleton, G.W.R. Director 1878-1890, Deputy Chairman until he resigned 1891.
Cobham Hall	Four miles SSE of Gravesend, Kent. Once the home of Viscount Cobham and now a school. See *Cobham.*
Cockington Manor	Domesday Manor but now known as Cockington Court near the model village west of Torquay, South Devon. Now owned by the Devon Rural Skills Trust.
Coeur de Lion (2)	Popular name for *King Richard I*, the Lion Heart. In this context, subject of a novel by Sir Walter Scott.
Cogan Hall	Cogan Old Hall, now a farm-house in Sully Road, Penarth, S. Glamorgan.
Coity Castle	Just E of Bridgend in Mid Glamorgan.
Colombo	When visited on the Royal Cruise 1901 (see text) the island of which it is the capital was known as Ceylon, now Sri Lanka.
Colonel Edgcumbe	G.W.R. Director, 1899-1915.
Colston Hall	Public building in Bristol.
Columbia (2)	Nos. 3453-71 (later 3391-3409) were all obviously named after places in the British Empire and British Columbia must have been intended. It has been suggested that the name was too long for the standard nameplate but this cannot be right as *Dominion of Canada*, the first of the same batch, had a longer name. The T.V. example was built in Philadelphia and one of the many American Columbias was in the same State.

Comet (4)	By context, the G.W. and S.D. examples were "hazy objects with tail moving in a path about the sun", evocative of speed. The T.V. context gives no clue but is assumed to be the same subject.
Commodore	Commander of a squadron of ships.
Compton Castle	Fortified manor house at Marldon, Paignton, S Devon.
Compton Manor	Alternative name for *Compton Castle* and mentioned in Domesday Book. Swindon records show that they regarded these as different buildings. (Fig. P83).
Condover Hall	Four miles S of Shrewsbury. Now a school for the blind.
Coney Hall	The original building, at Colletts Green Road, Powick, Worcester was demolished in 1862 and since replaced by a detached house which uses the same name.
Conqueror	One of the British Commanders in the Boer War, 1899-1902. The name was allotted but not used, see correction to Part 7, page G36, see text.
Conyngham Hall	At Knaresborough, N Yorkshire.
Cookham Manor	The ancient Manor of Cookham, three miles N of Maidenhead, is an area of open land. There never has been a manor house so named although Lullebrook Manor stands in Cookham village. The National Trust has been Lord of the Manor of Cookham since 1934.
Copshaw	Name given by contractor who owned the locomotive after its L.&M.M. days. Also given as *Copshawe*, see text.
Coquette	French equivalent of *Flirt*, see text.
Corfe Castle	Fortress near Swanage, Dorset, which gives its name to the area.
Cormorant	Large dark-greenish diving bird, common in Britain.
Corndean Hall	Four miles NE of Cheltenham, Gloucestershire.
Cornishman	A batch with names drawn from both Devon and Cornwall should have had a "Devonian" as well as a Cornishman?
Cornubia (2)	Roman name for Cornwall (There *was* a *Devonia*, but not in the same context, see later.)
Cornwall	Sir Morton Peto, contractor for the construction of the Cornwall Minerals Railway (and an ancestor of the principal compiler of these notes) is believed to have had a locomotive so named, later employed by Wm. West & Sons.
Corsair (2)	Pirate or pirate ship. A fitting twin for *Brigand*, which see.
Corsham Court	Four miles SW of Chippenham, Wiltshire.
Cory Hall	Once a temperance Memorial Hall in Cardiff, but now demolished.
Coryndon (2)	Name given by the contractor who worked the Hayle Railway and which he used on two locomotives and on another he used elsewhere.
Cossack	One of a people of Turkish origin inhabiting parts of SE Russia, famous as horsemen.
Cotswold	Part of Gloucestershire and Oxfordshire which includes the Cotswold Hills.
Countess	See *The Countess.*
Countess Vane	Wife of Earl Vane, Director of the Newtown & Machynlleth Railway, later part of the Cambrian Railways.
County . . .	Ten of the 4-4-0 "Counties" were named after counties in SE Ireland, which did not use the "County of . . ." form.
County of . . .	English and Welsh counties, mostly on the G.W.R. system, used on the 4-4-0's and 4-6-0's (with minor variations). The County names were those which applied before the 1974 Local Government changes.
Courier (3)	Express messenger. In this context, evocative of speed of the locomotive. Just possibly adopted from the name of a ship or stage coach, see text.
Cranbourne Grange	Private house at Sutton Scotney, Hampshire.

Cranbrook Castle (3) Iron Age encampment one mile S of Drewsteignton, N Devon.
Cranmore Hall East Cranmore, four miles E of Shepton Mallet, Somerset.
Cransley Hall Three miles SW of Northampton.
Crawley Grange Near Newport Pagnell, Buckinghamshire. Now converted into flats.
Creese Malayan dagger. See text.
Creon In mythology, a ruler and member of a Royal House.
Criccieth Castle Dominates a hill-top west of the seaside town of Criccieth, Gwynedd.
Cricket Perhaps evocative of the agility of a shunting locomotive, see text.
Crimea Black Sea peninsula, site of war with Russia, 1853-6.
Cromwell's Castle Built in 1651 in the Isle of Tresco to replace King Charles' Castle.
Croome Court Stately home eight miles E of Great Malvern in Hereford & Worcester which was prepared as an emergency home for the King and Queen in the event of a German invasion in 1940. Once known as Crome Court.
Crosby Hall No location given in Swindon records but most likely to be the hall so named built in Bishopsgate, London, in 1466. It was dismantled brick by brick and re-erected in Chelsea in 1909.
Crosswood Hall Ten miles SE of Aberystwyth, Dyfed. Now headquarters of the Welsh Agricultural Development & Advisory Service.
Crow The popular name for the Carrion Crow, a common British bird.
Croxteth Hall West Kirby, Liverpool. Now a country park museum.
Cruckton Hall Four miles WSW of Shrewsbury. Now a school.
Crumlin Hall Four miles SW of Pontypool, Gwent. Now the College of Mining.
Crusader Five of the 3001-40 batch had names connected with wars of various kinds.
Crynant Grange Monastic Grange six miles N of Neath, attached to *Neath Abbey*.
Cupid Roman God of Love.
Cwmcarn The purchasers of G.W. No. 692 had a colliery at Risca, Gwent, known as Cwmcarn Colliery.
Cwm Mawr (2) Northern terminus of the Burry Port & Gwendraeth Valley Railway.
Cyclops (7) One-eyed giant in Greek mythology. The name means "Round-Eye".
Cyfarthfa May refer to Cyfarthfa Ironworks in Merthyr or the Guest family home on the outskirts of the town. See *Sir Ivor* and *Lady Cornelia*.
Cyfronydd Residence of R. D. Pryce, Director of Newtown & Machynlleth Railway.
Cymbeline Ancient English King and father of Caractacus.
Cymmer (2) Colliery three miles WNW of Pontypridd, served by Taff Vale Railway.
Cynon Tributary of the Taff, which it joins at Abercynon.
Cyprus In context, another name of Aphrodite, the Greek Goddess of Love, and not the Mediterranean Island.
Czar Title of the former Emperors of Russia.

Daisy (2) The A.D. locomotive was named by a previous owner; the wartime name was one of a radio dual act, "Gert and Daisy".
Dalton Hall Swindon records it six miles NW of Beverley, Humberside.
Damon (2) Spelt Dameon in Greek mythology. Son of Phlius and ally of Heracles.
Danish Monarch Renamed to avoid confusion with "King" Class locomotives, see text.
Daphne Name given by the Metropolitan Railway, see text.
Dare Tributary of the River Cynon (which see), with confluence at Aberdare.

Dart (4)	The "Fire Fly" Class locomotive was twinned with *Arrow* and is clearly a "small pointed missile"; the B.&M. example had been named by a previous owner; the G.W. example was the Devonshire river which might be expected to have inspired the naming of the S.D. example but the makers, Slaughter, Gruning & Co., also built the two warlike "Fireflies", see above.
Dartington Hall	Two miles N of Totnes, South Devon. Now a school.
Dartmoor	Wild area of North Devon.
Dartmouth	South Devon sea-port, famed for its naval college.
Dartmouth Castle	15th century castle guarding the Dart estuary.
Davenham Hall	Two miles S of Northwich, Cheshire. Now a nursing home.
David MacIver	G.W.R. Director 1875-1907. Son of one of the two founders of the Cunard Line. Lived at Birkenhead.
Davies	One of the two contractors who worked the Pembroke and Tenby Railway until 1870.
Dean	See *William Dean.*
Dee (3)	River in N.Wales and Cheshire.
Defiant	World War II night-fighter, allied to the Spitfire, which see. (Fig. P88).
Denbigh Castle (4)	Built in 1282 at Denbigh, now a Clwyd town.
Derby	J. T. Firbank built the Derby-Eggington Junction Railway in 1875-8.
Derwent Grange	Private home near Oswestry in Shropshire.
Devizes Castle (3)	Still a residence at Devizes in Wiltshire.
Devonia	Part of North Devon incorporating Barnstaple, South Molton and Lynmouth.
De Winton	William de Winton, first Chairman of the Brecon & Merthyr Railway.
Dewrance	John Dewrance (died 1861). Helped in building of George Stephenson's Rocket and later was locomotive superintendent of L.&M. Railway.
Diamond	Name given by a former owner. perhaps an allusion to "Black Diamonds", a euphemism for coal.
Diana	Mythical Roman woodland Goddess and mythical huntress.
Dickinson	Probably a Director or Officer of the Ebbw Vale Steel, Iron & Coal Co.
Didlington Hall	Stood at Thetford in Norfolk until demolished about 1950.
Dido (2)	Legendary Queen of Carthage in Greek mythology.
Dinas (2)	Colliery served by the T.V.R. near what now is Dinas Rhondda station.
Dingley Hall	Two miles E of Market Harborough, Leicestershire.
Dinmore Manor	Eight miles N of Hereford in Hereford & Worcester.
Dinton Hall	Three miles SW of Aylesbury, Buckinghamshire.
Disraeli	Benjamin Disraeli, (Earl of) *Beaconsfield.* Prime Minister 1868 and 1874-1880.
Ditcheat Manor	Between Shepton Mallet and Castle Cary in Somerset.
Djerid	Oriental Dagger, see text.
Dodington Hall	Eight miles WNW of Bridgwater, Somerset.
Dog Star	Popular name for *Sirius*, the brightest star in the sky.
Doldowlod Hall	Known locally as Doldowlod House, five miles SSE of Rhayader, Powys.
Dolhywel Grange	Monastic Grange at Trecastle, Powys, attached to Talley Abbey.
Dominion of Canada	Part of the British Empire when the locomotive was named.
Donnington Castle	Two miles N of Newbury, Berkshire.
Donnington Hall	Three miles S of Ledbury, Hereford & Worcester.
Dorchester Castle	In this Dorset town, now used by the County Regiment.
Dorford Hall	One mile W of Nantwich, Cheshire. Known locally as Dorfold Hall.
Dorney Court	One mile NW of Windsor, Berkshire.
Dorothy (2)	Both locomotives had Powlesland & Mason connections, see text.

Dot Name given by I. W. Boulton.
Dove Named by contractor who built part of Cambrian Railways.
Dowlais (2) Suburb of Merthyr served by a Taff Vale branch.
Downham Hall Four miles NE of Clitheroe, Lancashire.
Downton Hall Six miles SE of Craven Arms, Shropshire.
Dragon (4) Mythical fire-breathing monster, obviously evocative of steam locomotives.
Draycott Manor Four miles NNE of Chippenham, Wiltshire, known locally as Draycot Manor.
Dreadnought Name evocative of the sheer power of the steam locomotive.
Dromedary (2) Beast of burden, evocative of the work of the steam locomotive.
Druid (2) Celtic priest. The Birkenhead Railway renamed *Birkenhead* as *Monk* to be in keeping with this choice. The G.W. example was one of eight miscellaneous names apparently chosen at random.
Drysllwyn Four miles W of Llandeilo on the Llanelly Railway's Carmarthen branch.
Drysllwyn Castle (3) Ruined castle dominating a mound at *Drysllwyn* (see above).
Duchess of Albany Wife of Queen Victoria's son Leopold.
Duchess of Teck Daughter of 19th century Duke of Cambridge and mother of *Princess May.*
Dudley Castle Ruined 13th century castle in West Midlands.
Duffryn Almost certainly Middle Duffryn on the Taff Vale's Aberdare line.
Duke of Cambridge Queen Victoria's cousin who died in 1904.
Duke of Connaught Queen Victoria's son, *Arthur*, born 1850.
Duke of Cornwall Title given to the eldest son of the reigning monarch.
Duke of Edinburgh Queen Victoria's son, *Alfred*, born 1844.
Duke of Lancaster Name given by Mersey Railway.
Duke of York (2) King Edward VII's son, George, who succeeded his father in 1910.
Dumbleton Hall Six miles SSW of Evesham, Hereford & Worcester.
Dummer Grange Five miles SW of Basingstoke, Hampshire.
Dunedin Port on New Zealand's S Island, visited on Royal Cruise, 1901, see text.
Dunley Hall Five miles S of Bewdley, Hereford & Worcester.
Dunley Manor Four miles N of Whitchurch, Hampshire.
Dunraven Colliery at Treherbert, opened at the time the T.V. reached there.
Dunraven Castle Fortified manor house built in the 11th century on the site of the house of Saxon leader Caractacus five miles S of Bridgend in Mid Glamorgan. Once the home of the *Earl of Dunraven.*
Dunster Castle On high land at Dunster, N. Somerset. Ancestral home of G.W.R. Director Geoffrey Luttrell.
Dunvant On the Llanelly Railway's Pontardulais-Swansea section.
Durban Capital of South Africa, visited on Royal Cruise 1901, see text.
Dusty Alternatively quoted as *Duty*, name given by colliery company.
Dutch Monarch Renamed to avoid confusion with "King" Class locomotives, see text.
Dwarf Name given by contractor engaged in building part of Cambrian Railways' system.
Dymock Grange Originally a Monastic grange attached to Flaxley Abbey, now a private residence known as Dymock Old Grange, Dymock, Gloucestershire.
Dynevor Castle A princely castle on a lofty ridge one mile W of Llandeilo in Dyfed. For many years the home of Lord Dynevor. A new castle now stands just north of the ruins of the old one.
Dyvatty E of Burry Port on the B.P.&G.V.'s Llanelly branch.

Eadweade Saxon for Edward, son of King Alfred.
Eagle (2) This large bird is symbolic of the power of a steam locomotive.
Eahlswith Wife of King Alfred.

Earl Baldwin	See *Stanley Baldwin.*
Earl Bathurst (2)	of Cirencester Park, Cirencester, Gloucestershire.
Earl Cairns (2)	of Farleigh House, Farleigh Hungerford, Bath, when the name was given.
Earl Cawdor (3)	Formerly (Viscount) *Emlyn.* G.W.R. Director from 1890, Deputy-Chairman from 1891 and Chairman 1895 until appointed First Lord of the Admiralty in 1905. Then lived at *Stackpole Court,* Dyfed. His descendant was honoured on the "Dukedog" and the "Castle".
Earl of Berkeley (2)	Owner of *Berkeley Castle,* Gloucestershire. The title became extinct on his death in 1941.
Earl of Birkenhead	Lived at Banbury, Oxfordshire.The title became extinct on the death of the Earl's son in 1984.
Earl of Chester (2)	Title bestowed by Queen Victoria on her eldest son, *Albert Edward,* later *King Edward VII.*
Earl of Clancarty	Lived in the Oxford area.
Earl of Cork	G.W.R. Director 1890-1904, who lived near Frome, Somerset.
Earl of Dartmouth (2)	Lord Great Chamberlain of England, 1928-36 and a G.W.R. Director, whose country seat was *Patshull Hall,* which see.
Earl of Devon (3)	Chairman of Bristol and Exeter Railway when it was amalgamated with the G.W.R. in 1876 and he and his descendants joined G.W.R. Board. Owned *Powderham Castle* and *Holker Hall,* which see.
Earl of Ducie (2)	Lived at *Tortworth Court,* Falfield, Gloucestershire.
Earl of Dudley (2)	G.W.R. Director 1936-1947. Lived at *Himley Hall* near Dudley.
Earl of Dunraven (2)	G.W.R. Director 1925-1935. Once owned *Dunraven Castle,* which see.
Earl of Eldon (2)	Lived at Winchester, Hampshire.
Earl of Mount Edgcumbe (2)	G.W.R. Director 1923-1943. Family name *St. Aubyn* carried on a "Bulldog". Lived at *Mount Edgcumbe* near Plymouth.
Earl of Plymouth (2)	G.W.R. Director, 1922-1923, representing the Barry Railway and whose home was *St. Fagans Castle.*
Earl of Powis	Lived at *Powis Castle,* Welshpool, Powys.
Earl of Radnor (2)	Lived at Longford Castle, Salisbury, Wiltshire.
Earl of St. Germans (2)	Lived at Port Eliot, St. Germans,Cornwall.
Earl of Shaftesbury	Lived at Wimborne, Dorset. One of his ancestors founded the Shaftesbury Society and the statue of Eros in London is a memorial to his work for poor children.
Earl of Warwick	The Earl of Warwick and Brooke owned *Warwick Castle* and was A.D.C. to Queen Victoria.
Earl St. Aldwyn	Lived at Williamstrip Park, Gloucester.
Earl Waldegrave	Lived at Bath, Avon.
Eastbury Grange	Two miles ENE of Lambourn, Berkshire. Now a farm.
Eastcote Hall	Two miles E of Solihull, West Midlands.
Eastham Grange	Three miles E of Tenbury Wells, Hereford & Worcester.
Eastnor Castle (3)	A noted tourist attraction one mile ESE of Ledbury, Hereford & Worcester.
Easton Court	Five miles SE of Ludlow, Hereford & Worcester.
Easton Hall	Stood near Grantham, Lincolnshire until demolished in 1951.
Eaton Hall	Stood two miles N of Congleton, Cheshire until demolished in 1981 and not to be confused with the Duke of Westminster's residence.
Eaton Mascot Hall	Six miles SE of Shrewsbury, Shropshire and known locally as Eaton Mascott Hall.
Ebor	Shortened form of "Eborarcum", the Roman name for York.
Eclipse (2)	An appropriate name for one of the early B.G. "Sun" Class. Several of the 3320-39 batch of "Bulldogs" repeated early B.G. names.
Eddystone	Rock 15 miles SW of Plymouth on which *Smeaton* built a lighthouse.

Eden Hall	Stood near Penrith, Cumbria, until demolished in 1931.
Edgcumbe	See *Colonel Edgcumbe.*
Edinburgh	See *Duke of Edinburgh.*
Edith Mary	Name given by a previous owner.
Edstone Hall	Five miles NNW of Stratford-upon-Avon, Warwickshire.
Edward VII	King of Great Britain, 1901-1910.
Edward Thomas	Manager and later Director of the Talyllyn Railway.
E. H. Llewellyn	Variant of *Evan Llewellyn* (which see), possibly carried for a short time.
E. J. Robertson Grant	Married a sister of the Waddell family (who built and worked the L.&M.M.R.) and who is believed to have held the road speed record, Lands End to John o' Groats.
Electra	Daughter of Oceanus and mother of the Goddess *Iris* (Mythology).
Elephant (3)	The largest existing mammal was certainly evocative of the power of the steam locomotive.
Elfin	Evocative name for a small shunting locomotive.
Elk (3)	A fleet of foot animal evocative of the speed of a locomotive.
Elmdon Hall	Demolished in 1948 to make way for the building of Birmingham Airport.
Elmley Castle	Little now remains of this large castle, sited four miles SW of Pershore, Hereford & Worcester.
Elton Hall	Five miles NE of Oundle, Northamptonshire.
Ely	River discharging into the Bristol Channel at Penarth.
Emlyn (3)	G.W. Nos. 3041/71 were named after Viscount Emlyn, another title of *Earl Cawdor*, which see. C. D. Phillips of Emlyn Works, Newport, who supplied the W.C.&P. locomotive applied the name Emlyn, usually accompanied by a number, in his sales catalogues.
Emperor (4)	Usually the title of the supreme ruler of an empire; in this context may have been taken from the name of a ship built in 1843.
Empire of India	Part of the British Empire when the locomotive was named.
Empress of India	One of Queen Victoria's titles.
Enborne Grange	Three miles SW of Newbury, Berkshire. Now a hotel.
Enterprise	Name given by contractor who worked the Llanidloes & Newtown Railway.
Enville Hall	Ten miles SW of Wolverhampton, W Midlands.
Erebus	Mythical son of Chaos; also the darkest depth of the Underworld.
Erlestoke Manor	Six miles SW of Devizes, Wiltshire. Demolished except for gateposts.
Ernest	Prince Alfred Ernest was Queen Victoria's second son. The name *Alfred* had been used already (see earlier) and the choice is curious as he was the *Duke of Edinburgh* and the next locomotive acquired by the Llanelly Railway was named *Edinburgh*, which see.
Ernest Cunard	G.W.R. Director 1907-22. Grandson of one of the founders of the Cunard shipping line.
Ernest Palmer	See *Lord Palmer.*
Eshton Hall	Five miles NW of Skipton, N. Yorks. Dates back to 12th century but rebuilt 1825/6. Now a nursing home.
Esk	Cumbrian river which discharges into the Solway Firth, see text.
Estaffete	Wrong spelling of French Estafette, a military courier, evocative of speed.
Estevarney Grange	Monastic grange at Monkswood, two miles NW of Usk, Gwent, and attached to *Tintern Abbey.* Known locally as Estavarney Grange.
Ethon	Mythical eagle or vulture which gnawed Prometheus' liver.
Etna (3)	Sicilian volcano. In context, probably evocative of steam and smoke.

Etona — Latin equivalent of Eton. Almost certainly chosen by Neville Grenville, see text, as he was a scholar here.

Eupatoria (3) — This batch of B.G. locomotives had names linked to the Crimean War but it is not clear why this name out of the seven originally used should have been put on No. 3078 in 1906.

Euripedes — Athenian tragic poet.

Europa — Mythical daughter of King Agenor, loved by the Greek God Zeus.

Evan Llewellyn — G.W.R. Director 1898-1914 who served in the Boer War, 1899-1902. The locomotive may have been named *E. H. Llewellyn* at first, which see.

Evening Star (2) — Poetic and popular name for *Venus* (see also *Morning Star*). Name also used for last steam locomotive built at Swindon, B.R. No. 92220.

Evenley Hall — One mile S of Northampton. Now a children's home.

Evesham Abbey — Pre-Norman Conquest Abbey at Evesham, Hereford & Worcester.

Excalibur — King Arthur's sword in the Arthurian legends.

Exe (2) — River which rises on Exmoor and joins the English Channel at Exmouth.

Exeter — County town of Devonshire. Renamed *Smeaton* when *City of Exeter* was built (which see).

Exmoor — Somerset moorland famous for its wild ponies.

Eydon Hall — Eight miles SSW of Daventry, Northamptonshire.

Eynsham Hall — Six miles NW of Oxford. Once the home of G.W.R. Director James Mason.

Eyton Hall — Location not given in Swindon records. There is one at Ruabon, Clwyd, and another at Leominster, Hereford & Worcester and at least another six elsewhere.

Faendre Hall — Four miles NE of Cardiff, S Glamorgan.

Fagan — Unofficial name given in Second World War, see text.

Fairey Battle — Fighter-bomber used by the R.A.F. early in World War II.

Fairfield — Area of Bow, E London, where the B.&E. Steam Railmotor was built.

Fairleigh Hall — Probably Farleigh House, four miles W of Trowbridge, Wiltshire, and once the home of *Earl Cairns*.

Fair Rosamund — Probably Rosamund de Clifford of *Clifford Castle* (which see), who lived in a bower near Woodstock, Oxfordshire, built by her Royal lover, King Henry II.

Fairy — Name given by a contractor and hardly suitable for a steam locomotive.

Falcon (2) — Bird of prey with outstanding powers of flight, evocative of the steam locomotive.

Falmouth (2) — Terminus of West Cornwall Railway branch from Truro, name used again in a West Country context on a "Duke" Class locomotive.

Farleigh Castle — Farleigh Hungerford Castle stands on a falling spur four miles W of Trowbridge, Wiltshire. An old manor house, crenellated in 1383 and now a ruin.

Farnborough Hall — Six miles NW of Banbury, Oxfordshire.

Farnley Hall — Two miles NW of Otley, N Yorkshire.

Fawley Court — One mile N of Henley-on-Thames, Oxfordshire. Now a school.

Fenton — James Fenton, 1815-63. Locomotive engineer, assistant to *Brunel*.

Fillongley Hall — Six miles NW of Coventry, W Midlands.

Fire Ball . . . . In company with *Fire Brand*, *Fire Fly* and *Fire King* were evocative names probably chosen by the makers of an early standard design. See also *Spit Fire* and *Wild Fire*.

Firefly — Name given by L.&N.W.R. after acquisition from Birkenhead Railway.

Fire King	The early name, see above, was used again on No. 3010.
Flamingo	A colourful but perhaps unsuitable name for a steam locomotive.
Fledborough Hall	Once stood at Holyport, one mile E of Maidenhead, Berkshire.
Fleetwood	The contractor Thos. Riley, an earlier owner, was based in Fleetwood.
Fleur-de-Lis	Former Royal Arms of France but confused with the Prince of Wales' feathers, see text.
Flirt	A singularly unsuitable name for a locomotive, see text.
Flockton Flyer	Fictional name given while used in a children's TV programme.
Flora	Mythical Goddess of flowers.
Florence	Either a girl's name or the Italian city, but see text.
Flying Dutchman	Racehorse, winner of the Derby in 1849, see text.
Flying Fortress	Informal name given to a W.D. "Dean Goods" locomotive in World War II. United States Air Force heavy bomber.
Foremarke Hall	Seven miles S of Derby, known locally as Foremark Hall. Now a school.
Forerunner	There is no obvious reason for so naming a very ordinary 0-6-0.
Forester (2)	Mediaeval Royal Officer of the Forest of Dean, who looked after the "vert" (vegetation) and venison; a paid office.
Forth	River on which Edinburgh stands (see text).
Forthampton Grange	Monastic grange two miles W of Tewkesbury. Attached to Tewkesbury Abbey.
Foster	Locomotive engineer 1775-1860 who helped to build "Puffing Billy".
Fountains Hall	Built from some of the stone from nearby Fountains Abbey, near Ripon, N Yorkshire. Now owned by North Riding County Council.
Fowey (2)	China clay port, southern terminus of Cornwall Minerals Railway.
Fowey Castle	At Fowey but known as St. Catherine's Castle. Built by Henry VIII.
Fowey Hall	At Fowey, Cornwall.
Fowler	Sir John Fowler, appointed Consulting Engineer to G.W.R. from 13/12/1860 in succession to I. K. Brunel.
Fox	The W.C.R. gave this name to a very small shunting locomotive, perhaps evocative of its agility. The A.D. name had been given by the Mersey Railway, possibly someone connected with the line.
Foxcote Manor	One mile SW of Andoversford, Gloucestershire.
Francis	Temporary informal World War II name, see text.
Francis Mildmay	See *Lord Mildmay of Flete*.
Frank Bibby	G.W.R. Director 1897-1923. Racehorse owner who won the Grand National in 1907 with *Kirkland* and 1911 with *Glenside*.
Frankton Grange	Three miles SW of Ellesmere, Shropshire.
Frederick Saunders	G.W.R. Director from 1886, Chairman 1889-95.
Frensham Hall	Five miles S of Farnham, Surrey. Once the home of the Dowager Countess Cawdor, wife of G.W.R. Chairman *Earl Cawdor.*
Freshford Manor	Five miles SE of Bath, Somerset.
Frewin Hall	Attached to *Brasenose* College, Oxford.
Friar Tuck	Character in the Robin Hood legends.
Frilford Grange	Three miles W of Abingdon, Oxfordshire.
Frilsham Manor	Six miles NE of Newbury, Berkshire.
Fringford Manor	Domesday manor four miles NNE of Bicester, Oxfordshire.
Fritwell Manor	Domesday manor six miles N of Bicester, Oxfordshire.
Fron Hall	Two miles SE of Mold, Clwyd.
Fulton	Hamilton Henry Fulton, 1813-86. S London railway engineer.
Fury	Mythical female spirit of justice and vengeance.

——oOo——

Gadlys	Site of colliery near Aberdare, served by T.V.R.
Gallo	Iberian name probably given on an earlier contract in Buenos Aires.
Ganymede	Cup bearer to Zeus in Greek mythology.
Gardenia	Sub-tropical shrub with white or yellow flowers.
Garsington Manor	Stands in a SE suburb of Oxford.
Garth Hall	Not identified in Swindon records. There are a number of Halls so named.
Gatacre Hall	Five miles ESE of Bridgnorth, Shropshire.
Gaveller	Crown officer in charge of mining in the Forest of Dean.
Gazelle (5)	Small, swift antelope, evocative of the agility of steam locomotives.
Gelly Gaer	T.V. names of the period were often taken from branch lines but Gelly Gaer as such was not rail-connected.
Gemini	3rd sign of the Zodiac. The Twins.
General Don	Name given by a contractor.
General Wood	Appears to have connections with the M.&M.R., see text.
George A Wills (2)	G.W.R. Director 1911-1927 in place of his cousin, Lord *Winterstoke*. Chairman of the Imperial Tobacco Company of Bristol.
George Waddell	One of the family who built and worked the L.&M.M.R.
Gert	Temporary informal World War II name, see *Daisy*.
Geryon	Mythical three-headed and three-bodied monster.
Gheber	Follower of an ancient Persian religion.
Giant	Name given by the L.&N.W.R.
Giaour	Turkish name for a Christian; the context gives no clue to this strange choice.
Gibraltar	Former British naval base at the W end of the Mediterranean, visited on Royal Cruise, 1901, see text.
Gipsy	Name given by Thomas Brassey, locomotive contractor to N.A.&H.R.
Gipsy Hill	Station on the L.B.&S.C. system when the locomotive was built.
Giraffe (2)	Animal possibly evocative of the tall locomotive chimneys common at the time. Not very evocative of locomotive speed!
G. J. Churchward	Chief Mechanical Engineer 1902-1921. (See also Fig. M7).
Gladiator (2)	Three members of the B.G. Class had warlike names and the Fleet Air Arm aircraft of World War II was appropriately named.
Gladstone	William Gladstone, 1809-1898. Liberal Prime Minister 1868-1874 and three later periods.
Gladys	Daughter of Earl Vane, Chairman of Newtown and Machynlleth Railway.
Glandovey	Cambrian Railway station between Machynlleth & Aberystwyth.
Glanmor	Iron foundry at Llanelli, Dyfed.
Glansevern (2)	Near Berriew, Welshpool. Residence of Mrs. Ann Owen, a major shareholder who performed the opening ceremony of the Llanidloes & Newtown Railway.
Glasfryn Hall	Swindon records location as a non-existent "Caerys", probably in error for Caerwys, near Ruthin, Clwyd. There are two such Halls, each about 12½ miles S and SW of Caerwys.
Glastonbury	Abbey town on Sedgemoor in Somerset. See also *Butleigh Court*.
Glastonbury Abbey	Ruined Abbey in the Somerset town of this name.
Glendower	Owen Glendower, 1350-1415. Welsh Chief who conducted guerilla war against the English on the border between England and Wales.
Glenside	Winner of the Grand National Steeplechase in 1911, owned by G.W.R. Director *Frank Bibby*. The horse was blind in one eye.
Gloucester (2)	The B.G. "Swindon" Class bore names of the principal places served by the G.W.R. The locomotive acquired by the T.V.R. had been named after the same city, as terminus of the Birmingham & Gloucester Railway.
Gloucester Castle	Once stood in the City of Gloucester.

Glyn	Name given by the L.&N.W.R.
Glyncorrwg	Northern terminus of South Wales Mineral Railway.
Gnat	A noxious insect which hardly epitomises a steam locomotive.
Gnome	An apt name *Chester* was changed to *Gnome* for no obvious reason.
Goat	With *Owl* and *Weasel*, a curious choice for 0-4-0 well tanks.
Godolphin	Very old mansion five miles NW of Helston, Cornwall. Former home of the Earls of Godolphin, it is now a farmhouse.
Goldfinch	Songbird with gold and red patterned head.
Goliah (2)	Alternative spelling of *Goliath*, a Biblical strong man.
Goliath	An apt choice for a 2-8-0T used to haul passenger trains on a preserved railway, over steep gradients.
Gooch (3)	See *Sir Daniel Gooch* and Fig. M3.
Goodmoor Grange	In the Wyre Forest, W of Kidderminster, Hereford & Worcester.
Goodrich Castle	Adjoins the main Ross-on-Wye to Monmouth Road, Hereford & Worcester.
Goonbarrow	Terminus of Cornwall Mineral Railway's Goonbarrow Branch.
Gopsal Hall	Stood at Atherstone, Warwickshire, until demolished in 1951. In its later years was known as Gopsall Hall.
Gordon	Scottish general, murdered at Khartoum in 1885, during the Sudanese War.
Gorgon (2)	Mythical female sea creature.
Gossington Hall	Near Coaley Junction on the L.M.S. Gloucester to Bristol line.
Goytrey Hall	Five miles NE of Pontypool, Gwent. More recently known as Goytre Hall, which is the correct Welsh spelling.
Grantley Hall	Six miles W of Ripon, N Yorkshire.
Granville Manor	Haddenham, Bucks. Known locally as Grenville Manor.
Grasshopper	See *Cricket*, equally evocative of the agility of a shunting locomotive.
Graythwaite Hall	Five miles W of Appleby, Cumbria.
Great Britain (3)	Almost certainly the ship designed by Brunel, launched in 1846, see text, and Fig. P87.
Great Mountain	Waddell's colliery near Cross Hands, served by the L.&M.M.R.
Great Western (4)	Abbreviated name of the Company, see text.
Green Dragon	Heraldic crest adopted by the Earl of Pembroke & Montgomery.
Gresham Hall	Five miles SW of Cromer, Norfolk.
Greyhound (2)	Could have been a stagecoach name (see text) or a name generally evocative of speed.
Grierson	First G.W.R. General Manager, 1863-87.
Grongar	Location on the Llanelly Railway's Llandilo-Carmarthen line.
Grosvenor	Family name of the Duke of Westminster, to whom the Directors wished to pay a compliment.
Grotrian Hall	Once stood at 115-7, Wigmore Street, London, W1. Now demolished.
Grundisburgh Hall	Three miles WNW of Woodbridge, Suffolk.
Guernsey	See *Isle of Guernsey*.
Guild Hall	The City of London's official public hall.
Guinevere	King Arthur's Queen.
Guy Mannering	Character who gave his name to a novel by Sir Walter Scott.
Gwenddwr Grange	Monastic grange six miles SSE of Builth Wells, Powys, attached to Abbeydore.
Gwendraeth (2)	Dyfed river, S of Carmarthen, which parallels the B.P.&G.V.R.
Gyfeillon	Name of a colliery near Pontypridd on the Taff Vale Railway, bought by the G.W.R. in 1854 to ensure continuity of supplies when another source failed. The colliery was sold in 1865.

Haberfield Hall	At Pill, three miles NW of Bristol.
Hackness Hall	Four miles WNW of Scarborough, N Yorkshire.
Hackworth	Timothy Hackworth, 1786-1850, designer and builder of *Puffing Billy*.

Haddon Hall	Domesday manor two miles SE of Bakewell, Derbyshire, some parts of which were incorporated in the present house.
Hades	Mythical god of the Underworld.
Hagley Hall	Two miles SE of Stourbridge, West Midlands. Once the seat of (Viscount) *Cobham*, G.W.R. Chairman 1889-91.
Halifax	Capital of Nova Scotia, Canada, visited on Royal Cruise, 1901, see text.
Hampden	R.A.F. bomber aircraft used in the early part of World War II.
Hampton Court	Although the Royal Palace in Middlesex is not on the G.W. system, no evidence has been found to link the name with another "Hampton Court" six miles SSE of Leominster, Hereford & Worcester.
Hanbury Hall	Three miles W of Droitwich, Hereford & Worcester.
Hanham Hall	No Swindon record. There is a Hanham House near Paulton, Avon and a Hanham Court near Bristol but either, or neither, could apply.
Hannington Hall	Five miles NNE of Swindon, Wiltshire.
Hardwick Grange	Monastic grange two miles S of Abergavenny, Gwent, attached to Abergavenny Priory.
Hare	This locomotive's origin is obscure, as is the reason for the name.
Harlech	Between Barmouth and Porthmadog on the Cambrian Railways coast line.
Harlech Castle	Dominates a promontory above Harlech Station.
Harold	Name given by a contractor when newly built.
Haroldstone Hall	At Broadhaven, near Haverfordwest, Dyfed.
Harpy	Mythical monster with a woman's head and body and birds' claws and wings.
Harriet	Possibly a relative of Christopher Rowlands.
Harrington Hall	Two miles ESE of Kidderminster, Hereford & Worcester, and known locally as Harvington Hall.
Hart Hall	Attached to Hertford College, Oxford.
Hartington	Marquis of Hartington, leader of the Liberal Party in Victorian days who held various Government offices between 1863 and 1874.
Hartlebury Castle	At Stourport in Hereford & Worcester. The official residence of the Bishops of Worcester.
Hatherley Hall	Three miles SW of Cheltenham,Gloucestershire. Demolished in the early 1970's and replaced by a block of flats with the same name.
Hatherton Hall	Three miles SE of Nantwich, Cheshire.
Haughton Grange	One mile N of Shifnal, Shropshire.
Haverfordwest Castle	Dominates the centre of this Dyfed town.
Haverhill	Named after an East Anglian town by the Colne Valley & Halstead Railway.
Hawk (3)	Bird of prey, the name evocative of speed.
Hawkstone	Name given by the L.&N.W.R.
Hawthorn	Founder of R.&W. Hawthorn & Co., locomotive builders of Newcastle.
Haydon Hall	Now a Council Youth Centre three miles NE of Uxbridge, Middlesex.
Hayle	Original western terminus of the Hayle Railway.
Hazeley Grange	At Cleobury Mortimer, Hereford & Worcester.
Hazel Hall	No Swindon record. The most likely subject is at Peaslake, Surrey, although there are other possibilities at Twyford, Berkshire, Coniston, Lancashire and Snainton, N Yorkshire.
Headbourne Grange	Two miles N of Winchester, Hampshire. Now in commercial use.
Heatherden Hall	Adjacent to Pinewood Film Studios at Iver Heath, Buckinghamshire.
Hebe	Mythical daughter of Zeus and cup-bearer to the Gods on Olympus.

Hecate (2)	Mythical Earth Goddess, associated with the World of the Dead.
Hecla (3)	Volcano in Iceland. A name evocative of smoke and steam.
Hector	Eldest son of King *Priam* of Troy.
Hecuba	Daughter of Dymas, King of the Phrygians.
Hedley	William Hedley, 1779-1843, who developed the idea of adhesion between locomotive wheels and running rails.
Helena	Daughter of Queen Victoria. See *Princess Helena.*
Helmingham Hall	16th century house at Debenham, near Stowmarket, Suffolk.
Helmster Hall	Near York. Known locally as Upper Helmsley Hall and had been built with materials from nearby Helmsley Castle.
Helperly Hall	Eleven miles NW of Harrogate, N Yorkshire and known locally as Helperby Hall.
Helston	Town eleven miles N of The Lizard, Cornwall.
Hengrave Hall	Embattled manor house replacing an older mansion four miles NW of Bury St. Edmunds, Suffolk.
Henley Hall	Two miles ENE of Ludlow, Shropshire.
Henshall Hall	Stood at Congleton, Cheshire, until demolished in 1958.
Hercules (9) } *Hercules No. 2* }	This mythical son of Zeus who performed feats requiring great strength was the most-used name of all.
Hereford (2)	Savin's locomotive was delivered to Hereford for use on the Hereford, Hay & Brecon Railway; the B.G. "Swindon" Class were named after the principal places on the G.W. system.
Hereford Castle	Stood in the City of Hereford.
Hero (4)	The two broad gauge examples were clearly the mythical priestess of Aphrodite who fell in love with *Leander*; the B.P.&P. name had been given by the St. Helens Railway; and the locomotive which hauled the test train across Whixall Moss (see page P66) had almost certainly been hired from the L.&N.W.R.
Heron (2)	This bird is a common sight on the River Towy at Carmarthen, which could have accounted for the C.&C.'s choice.
Herschell	Lord Herschell was a Victorian Lord Chancellor of England. One of his daughters married G.W.R. Director *Sir Arthur Yorke.*
Hesiod	Greek poet who lived about 800 B.C.
Hesperus (3)	Alternative name for the planet *Venus*, the *Evening Star.*
Heveningham Hall	Six miles SW of Halesworth, Suffolk, pronounced "Henningham".
Hewell Grange	Three miles NW of Redditch, Hereford & Worcester. Later became a remand home.
Highclere Castle	This stately home of the Earl of Carnarvon which stands near Newbury, Berkshire, is a magnificent mansion but hardly a castle.
Highnam Court	Three miles WNW of Gloucester.
Highnam Grange	Monastic grange three miles NW of Gloucester, attached to Gloucester Abbey.
Hilda	The L.&M.M. naming tradition suggests that she was a member of the Waddell family.
Hill	Sir James Hill, sometime Lord Mayor of Bradford, sponsoring authority of the Nidd Valley Waterworks.
Hillingdon Court	Two miles SE of Uxbridge, West London. Once the seat of Lord Hillingdon, G.W.R. Auditor, whose son *A. H. Mills* was a Director.
Himalaya	Name given by M.S.&L.R., one of a mountain range on N Indian border.
Himley Hall	At Dudley in Hereford & Worcester. Once owned by the *Earl of Dudley,* then by the Coal Board and then by the local Council.
Hinderton Hall	One mile E of Neston, Wirral, Cheshire. Once the residence of G.W.R. Director Sir Percy Bates, Bt.
Hindford Grange	Four miles NW of Oswestry, Shropshire.
Hindlip Hall	Two miles NNE of Worcester. Now a police training centre.
Hinton Manor	Seven miles W of Abingdon, Oxfordshire.
Hirondelle (4)	The French equivalent of the *Swallow*; in this context, evocative of locomotive speed.

H.M.B.	Initials of Hugh M. Bythway, industrialist connected with the Blaendare Colliery, Pontypool.
Hobart	Capital city of the Australian State of Tasmania, visited on Royal Cruise, 1901, see text.
Hodroyd	Name of colliery where G.W. No. 674 worked after sale.
Holbrooke Hall	Six miles N of Derby and known locally as Holbrook Hall.
Holker Hall	17th century stately home three miles SW of Cartmel, Cumbria, and the seat of the *Earl of Devon*, a G.W.R. Director.
Holkham Hall	18th Century Palladian Mansion at Holkham, N Norfolk.
Holmwood	Country house near Tunbridge Wells, Kent, owned by the Chairman of the Pembroke & Tenby Railway.
Homer	Greek poet who lived about 700 B.C.
Honington Hall	270 years old property two miles N of Shipston-on-Stour, Warwickshire.
Hook Norton	The locomotive worked for the Hook Norton Ironstone Partership Ltd., at Hook Norton, near Banbury.
Hook Norton Manor	Six miles SW of Banbury, Oxfordshire.
Hopton Grange	Monastic Grange five miles S of Montgomery, Powys, attached to Abbey Cwmhir.
Horace	Roman satirist and poet, 65-8 B.C.
Hornet	Another quite unsuitable name for a steam locomotive.
Horsley Hall	Seven miles SE of Market Drayton, Shropshire.
Horton Hall	Two miles NE of Chipping Sodbury, near Bristol.
Hotspur	The Earls of Percy and of March joined forces about 1400 to wage war against *King Henry IV* on the Welsh border. Percy, nicknamed Hotspur, was killed in action. G.W.R. Director *Charles Mortimer* was a descendant of the Earl of March.
Hove	The town west of Brighton was the source of this L.B.&S.C. name.
Howick Hall	Five miles NE of Alnwick, Northumberland. Pronounced "Hoik".
Hown Hall	At Taynton, seven miles WNW of Gloucester. The building is now in use as the farmhouse of Hownhall Farm.
Hubbard	See *Alexander Hubbard.*
Humber (2)	River separating N from S Humberside.
Hurricane (3)	The first *Hurricane* of October 1838 blew itself out in 14 months; the next use, in 1895, repeated the name of a racehorse; and the third example was named after the celebrated Second World War fighter aircraft.
Hurst Grange	Five miles E of Reading, Berkshire. Now a farm.
Hutton Hall	At Guisborough, Cleveland.
Hyacinth	Spring-flowering bulbs which come in many attractive varieties.
Hyacinthe	French version of the name, carried until 5/16.
Hydra	Mythical sea serpent with seven heads and the body of a hound.

Iago	This villain from Shakespeare's "Othello" sits strangely among three other members of this small class from Greek history and mythology.
Ickenham Hall	Now a youth centre, three miles NE of Uxbridge, West London.
Iford Manor	One mile SE of Bradford-on-Avon, Wiltshire.
Ilfracombe	Possibly named to advertise through G.W.R. services. The line from Barnstaple to Ilfracombe was owned and worked by the L.&S.W.R.
Ilkeston	Nine miles ENE of Derby and presumably connected with a contract.
Impney Hall	Near Droitwich, Hereford & Worcester. Now a hotel "Chateau Impney".
Ince Castle	Early 17th century castle overlooking the Lynher Estuary and opposite the village of Antony, eight miles W of Plymouth.
Inchcape	The Earl of Inchcape, G.W.R. Director 1918-32.
Inkerman	Site of battle in the Crimean War of 1853-6.

Inkermann	Misspelling of last name.
Inveravon	Supplied new to contractors Scott & Best and named after Mr. Best's Edinburgh residence.
Iris	In context, mythical goddess associated with the rainbow.
Iron Duke	Duke of Wellington, hero of Waterloo (but see text).
Ironsides	Possibly a play on words as the locomotive is reputed to have been built at Neath Abbey Ironworks (but not traced in their records).
Irthlingborough	Name given by the Ebbw Vale Steel, Iron & Coal Co. Ltd.
Isambard Kingdom Brunel	The first Engineer of the G.W.R., he was appointed in March 1833 and ended his career with the Saltash Bridge, opened in 1859.
Isebrook	Name given by a later purchaser; probably near Wellingborough.
Isis	Alternative name for the River Thames at Oxford.
Isle of Guernsey } *Isle of Jersey*	Originally *Guernsey* and *Jersey*, the original and expanded names were no doubt connected with the G.W.R.'s Channel Islands steamship services, taken over from another operator in 1889 and upgraded between 1897 and 1900.
Isle of Tresco	There were no G.W.R. steamers to the Scilly Isles and this name just had a West Country connection like many of its contemporaries.
Italian Monarch	Renamed to avoid confusion with "King" Class locomotives, see text.
Ivanhoe (2)	Wilfred of Ivanhoe, character in Sir Walter Scott's novel so named.
Ixion (2)	Mythical ancestor of the Centaurs and an unsavoury character.
Jackdaw	Britain's smallest member of the crow family.
Jamaica	West Indian Island, once part of the British Empire.
James Mason (2)	G.W.R. Director 1910-29. Lived at *Eynsham Hall.*
James Watt	Scottish engineer 1736-1819, who first realised the expansive power of steam. Partner in the engineering firm,Boulton & Watt.
Jane	Informal name given by Wantage Railway staff.
Janus	Roman God with two faces looking both ways, the origin of January which looks back to the Old Year and on to the New Year.
Japanese Monarch	Renamed to avoid confusion with "King" Class locomotives, see text.
Jason	Mythical hero of Thessaly, famed for his quest for the Golden Fleece.
Javelin	Short light spear (see text).
Jay (2)	A pinkish-brown crow, perhaps a suitable name for an 0-4-0ST.
J. C. Parkinson	Managing Director of the A.D.R.
Jean	Name given at Blaenavon Ironworks.
Jean Ann	Temporary informal name given in wartime, see text.
Jean Barbara	ditto.
Jeannie Waddell	Member of the Waddell family, see *George Waddell.*
Jersey	See *Isle of Jersey.*
Joan	Name given at Blaenavon Ironworks.
John	Name carried at Choppington Colliery, Northumberland.
John G. Griffiths (2)	G.W.R. Director 1908-22 and had been senior partner in the firm of Deloitte, Plender & Griffiths, auditors to the G.W.R.
John Gray	Locomotive superintendent of the Liverpool & Manchester Railway and later of the L.B.&S.C.R.
John Owen	Deputy Chairman of the Whitland & Cardigan Railway, which served his slate quarry.
John Waddell (2)	Head of the family firm of John Waddell and Sons, of Edinburgh, who were contractors to the Llanelly and Mynydd Mawr Railway.

John W. Wilson (2)	G.W.R. Director, 1908-32.
Joseph Shaw	G.W.R. Director, 1917-30.
J. R. Maclean	Not identifiable except that he was *not* a Director. He could have been a Consulting Engineer connected with building of the A.D.R.
Jumbo (2)	A famous London Zoo elephant of Victorian days was called Jumbo but an inappropriate name for the small Westlake and L.&S.W. locomotives.
Juno (2)	Mythical Roman Goddess of women and marriage.
Jupiter (4)	The G.W. examples were all named after the planet; the B.&M. locomotive was taken from the Roman "God above all other Gods".
Juvenal	Roman satirist, 60-140 A.D.
Kaiser	Emperor of Germany at the time the locomotive was built.
Kate	Name given by a different owner.
Katerfelto	A spirit said to flit across Exmoor on cloudy moonlight nights.
Keele Hall	Part of Keele University, Staffordshire.
Kekewich	General Sir George Kekewich, defender of Kimberley in the Boer War. Related by marriage to G.W.R. Director *Sir Massey Lopes.*
Kelham Hall	14th century home of the Manners family at Newark, Nottinghamshire and now a school.
Kenilworth	The context does not indicate whether the Warwickshire town or the Walter Scott novel was used as the source of this name.
Kenilworth Castle	Commenced in 1122, this once magnificent building is now a ruin. Queen Elizabeth I slept here for seventeen nights.
Kennet	Tributary of the Thames which it joins near Reading, Berkshire.
Kertch	Russian town with Crimean War associations.
Ketley Hall	Two miles ESE of Wellington, Shropshire.
Khan	Persian Governor or Asian Prince.
Khartoum	Capital of the Sudan, where (General) *Gordon* was murdered in 1885.
Kidwelly	Dyfed town served by the B.P.&G.V. and Gwendraeth Valleys Railways.
Kidwelly Castle	Dominates the Dyfed town of Kidwelly, eight miles S of Carmarthen.
Kilgerran Castle	Ruined castle outside the village of Cilgerran (Welsh spelling of Kilgerran), two miles S of Cardigan, Dyfed.
Killarney	*Ophir* was so renamed in connection with excursions from London to view the celebrated Lakes of Killarney.
Kilmar	Site of a quarry on the Liskeard & Caradon Railway.
Kimberley	Site of a battle in the Boer War (see *Kekewich*, above).
Kimberley Hall	18th century building at Wymondham, between Attleborough and Norwich in Norfolk. Little of the building now remains.
King	There is no obvious reason for so naming this small South Devon 2-4-0T.
King Arthur	Celtic warrior who fought against the Saxons. Leader of the Knights of the Round Table.
King . . .	Nos. 4021-30 were named after Kings of England in reverse order (with one exception, see text) starting from *King Edward* (VII) using each name only once and omitting the figure.
King . . .	Nos. 6000-29 were given the full titles of all the Kings of England in reverse order from *King George V. King Edward VIII* and *King George VI* were added later (see text page P99).
Kingfisher	The only species of Kingfisher found in Britain is the familiar brilliant blue, blue-green and chestnut fishing bird.
Kingsbridge	Devon town near Start Point, terminus of a closed G.W.R. branch.
Kingsland Grange	SW area of Shrewsbury, Shropshire. Now a school.
Kingsthorpe Hall	In N Northampton, owned by Local Authority.

Kingstone Grange	Monastic grange attached to Abbeydore and now a country house, at Thruxton, six miles SW of Hereford, Hereford & Worcester.
Kingsway Hall	Church premises in Kingsway, Central London. Derelict for some years.
Kingswear Castle	Opposite *Dartmouth Castle* across the River Dart, South Devon.
Kinlet Hall	Five miles NE of Cleobury Mortimer, Shropshire.
Kirby Hall	English Heritage property two miles SE of Cretton, Northamptonshire.
Kirkella	Kirk Ella, a village near Hull in Humberside.
Kirkland	Racehorse owned by G.W.R. Director *Frank Bibby* which won the Grand National in 1907.
Kitchener	Lord Kitchener, British Commander in the Sudanese and Boer Wars.
Kneller Hall	Near Twickenham, Middlesex. Built 1711 for SIr Godfrey Kneller, the painter, and now the home of the Royal Military School of Music.
Knight Commander	One of the classes of the Orders of Knighthood.
Knight of Liége	Belgian Order of Knighthood.
Knight of Liège	French Order of Knighthood.
Knight of St. John	Religious Order of St. John of Jerusalem.
Knight of St. Patrick	British Order of Chivalry fallen into disuse since 1922.
Knight of the Bath	British Order of Chivalry.
Knight of the Black Eagle	Prussian Order of Chivalry.
Knight of the Garter	Most Ancient and Most Noble Order of Chivalry of England.
Knight of the Golden Fleece	Austrian and Spanish Order of Chivalry.
Knight of the Grand Cross	One of the classes of some Orders of Chivalry, usually one of the highest.
Knight of the Thistle	British Order of Chivalry.
Knight Templar	Religious Order of Knighthood.
Knolton Hall	Seven miles S of Wrexham, Clwyd.
Knowsley Hall	At Prescot, E of Liverpool, Merseyside.

Lady Cornelia	Cornelia Henrietta Maria Guest, wife of *Sir Ivor*, which see.
Lady Cornewell	Name given by a Contractor.
Lady Disdain	Term addressed to Beatrice by Benedict in Shakespeare's "Much ado about nothing" but may be of even earlier origin.
Lady Elizabeth	Appears to have connections with the M.&M.R., see text.
Lady Godiva	Wife of Leofric, Earl of Mercia and mother of Hereward the Wake.
Lady Macbeth	Instigated the murder of King Duncan of Scotland in 1040.
Lady Margaret	Wife of Captain Spicer who had financed the L.&L. connection to the G.W.R. at Liskeard, opened in May 1901.
Lady Margaret Hall	Part of Oxford University.
Lady of Lynn	Title of a novel by Sir Walter Besant, 1836-1901.
Lady of Lyons	Pauline Descharpelles, heroine of a novel by Bulwer Lytton.
Lady of Provence	Character in a poem by Mrs. Hemans, said to have been founded on an incident in early French history.
Lady of Quality	Novel by Frances Hodgson Burnett who may have adopted the name from "The Memoirs of a Lady of Quality", published about 200 years ago.
Lady of Shalott	Arthurian lady, who died of unrequited love for *Sir Lancelot.*
Lady of the Lake	Deplorable damsel who was the mistress of *Merlin*, and having disposed of him, married Sir Pelleas.
Ladysmith	Town in Natal, S Africa, besieged by the Boers in the Boer War.
Lady Superior	The subject of Eliza F. Pollard's book so named, a "Jane Eyre" type novel published in 1875.

Lady Tredegar	Wife of *Lord Tredegar*, Chairman of the A.D. She cut the first sod for the A.D. Co.'s Alexandra Docks in May 1866.
La France	Name put on the first 4-4-2 De Glehn compound by its French makers.
Lagoon	Misspelling of Laocoon, a character in Greek mythology who warned the Trojans not to pull the wooden horse into Troy.
Laira	Site of the Plymouth locomotive shed, opened soon after No. 3338 (later 3326) was built.
Lalla Rookh (2)	Poem by Thomas Moore about a journey from Delhi to Kashmir by an Emperor's daughter, Lalla Rookh, see text.
Lambert	G.W.R. General Manager, 1887-96.
Lampeter	Market town at mid-point of Manchester & Milford Railway.
Lamphey Castle (3)	Built as a Bishop's Palace at Lamphey, Dyfed, crenellation in mediaeval times made it a Castle but it is still known as Lamphey Palace.
Lance (3)	This offensive weapon was chosen by Stothert & Slaughter (see text) for the "Sun" Class locomotive and the two S.D. examples were presumably equally lethal but the context is inconclusive.
Lanelay Hall	One mile W of Llantrisant, Mid Glamorgan, and now headquarters of Mid Glamorgan Fire Service.
Langford Court	Eleven miles SW of Bristol.
Langton Hall	Six miles WNW of Northallerton, N Yorkshire. Now a school.
Lark (2)	Popular name for the *Skylark*, appropriate for an 0-4-0ST.
La Savoie	Name of French "Departement" near the Swiss Border, where Mr. Brassey had a contract.
Latona (2)	Mythological Greek name for the mother of *Apollo* and *Diana*.
Launceston	County town of Cornwall.
Launceston Castle	Castle, entered in the Domesday Book, on the NW side of the Cornish capital. Now in a decayed condition.
Laurel	Evergreen shrub with dark glossy leaves.
Lawton Hall	No Swindon record. There is one three miles W of Leominster, Hereford & Worcester and another six miles N of Newcastle-under-Lyme, Cheshire.
Leander	Mythical lover of *Hero* who died swimming across the Hellespont.
Leaton Grange	Two miles W of Wellington, Shropshire.
Lechlade Manor	At Lechlade in SE Gloucestershire. Now a convent.
Leckhampton Hall	Stood at junction of Shurdington and Moorend Park Roads, Cheltenham, Gloucestershire. Demolished about 1958 and site redeveloped.
Leeds	Name apparently given by Liverpool & Manchester Railway.
Leighton	Village two miles SE of Welshpool, Powys. The context suggests the residence of someone connected with the Cambrian Railways but no facts have been discovered.
Leighton Hall	Two miles N of Carnforth, Lancashire.
Leo (2)	The B.G. locomotive was named after the 5th sign of the Zodiac, the Lion; the Llanelly example was Prince Leopold, one of Queen Victoria's sons.
Leon	Name given by a contractor, with a possible Spanish connection.
Leonidas	King of Sparta around 480 B.C.
Leopard (2)	Both examples were clearly evocative of the speed of a locomotive.
Leopold	King of Belgium when the locomotive was built.
Lethe	Mythological River of Forgetfulness in the Underworld.
Levens Hall	Six miles SW of Kendal, Cumbria. Now a private steam museum.
Libra	7th Sign of the Zodiac. The Balance.
Liddington Hall	Stood three miles SE of Swindon, Wilts. Completely demolished.
Liffey	The city of Dublin stands on this southern Irish river, see text.
Lightning (3)	Name evocative of speed and power of the locomotive.

TWO NAMES FOR THE PRICE OF ONE

Photograph] [K. Davies' Collection

No. 7807 was named *Compton Manor* **and Nos. 5047, 5072 and 5099 all bore the name** *Compton Castle* **at different times yet the Manor and the Castle, pictured above, are one and the same building!** P83

Photograph] [Charles Oldham

No. 6825 bore the anglicised name *Llanvair Grange* **. . .** P84

Photograph] [H. W. Robinson

. . . while No. 6877's nameplates carried the correct Welsh version *Llanfair Grange* **. . . but the same building was the inspiration for both names!** P85

WAR AND PEACE — 1

The Royal Navy cruiser *Powerful,* **whose guns were removed and hauled overland to help with the relief of Ladysmith in 1900 in one of the more spectacular episodes of the Boer War.**

Photograph] [Bill Peto's Collection

P86

Photograph] [ss Great Britain Project

Brunel's steamship *Great Britain* **was launched in 1843 and was in service for an exceptionally long period. The hulk survived to be returned to the special dock in Bristol where it was built and where it is being restored to its original condition. The picture is reproduced from a painting by Keith A. Griffin, showing ss** *Great Britain* **as it was in 1845.**

P87

WAR AND PEACE — 2

Boulton Paul Defiant F.Mk.I N1581 of No. 264 Squadron. (Dowty Photo.)

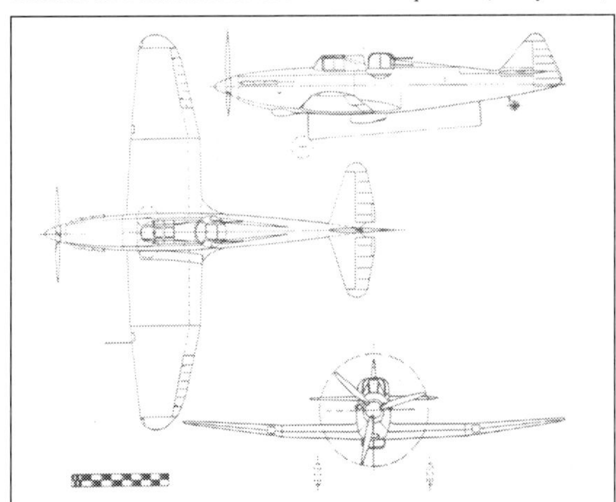

The Boulton Paul *Defiant* night fighter complete with its equivalent of a Swindon Works "Engine Diagram" and which gave its name to ....

Photograph] [Birmingham Railway Museum

. . . "Castle" Class No. 5080 *Ogmore Castle,* which was renamed *Defiant* in June 1941. Withdrawn in April 1963 and sold to Woodham Brothers, Barry, No. 5080 was rescued and restored to running order at Birmingham Railway Museum's Tyseley Depot in August 1987.

HISTORIC BUILDINGS AS LOCOMOTIVE NAMES

Photograph] [CADW Welsh Historic Monuments

Caerphilly Castle was chosen as the name for the prototype "Castle" Class 4-6-0 although the town of Caerphilly had only come into the G.W.R. fold when the Rhymney Railway became a constitutent company in 1922, just a year before No. 4073 was built. P89

Photograph] [K. Davies

The extensive Cistercian Monastery known as Neath Abbey was more than twice as long as Nos. 4063 and 5083 coupled together! P90

Lilford Hall	Just S of Peterborough, Cambridgeshire.
Lilleshall	Contractor's locomotive built by the Lilleshall Iron Co.
Lily	Species which includes the *Fleur-de-Lis*; in effect the name was duplicated in the same "1st Lot Metropolitan" 2-4-0WT.
Linden Hall	At Borwick, two miles NE of Carnforth, Lancashire.
Lion (2)	Both uses of the name were evocative of locomotive power and speed.
Liskeard	Northern terminus of Liskeard and Looe Railway.
Little John (2)	Character in the Robin Hood legends.
Little Linford Hall	Two miles W of Newport Pagnell, Buckinghamshire.
Littleton Hall	No Swindon record. The most likely is Littleton House, three miles NW of Winchester, Hampshire, but there are 28 on the maps in all!
Little Usk	The B.&M. had a 2-4-0 passenger locomotive and an 0-6-0ST both named *Usk* (which see). This nickname distinguished the 0-6-0ST.
Little Wyrley Hall	Four miles SE of Cannock, Staffordshire.
Liverpool	Name carried for only a few weeks, if at all, and presumably intended to "advertise" the G.W.R. service via Birkenhead, which was slower than that by the L.&N.W. route.
Lizard	See *The Lizard.*
Lizzie	Name given by a Contractor.
Llancaiach (2)	One of the Taff Vale's many branches was known as the Llancaiach Branch.
Llandaff (2)	Station on Taff Vale system, three miles N of Cardiff City Centre.
Llandinam	Named by Contractor, David Davies, who originally used it in the Llandinam area, two miles S of Moat Lane Junction.
Llandovery Castle	Remains of the castle stand SSW of this Dyfed town.
Llanerchydol	Hall near Welshpool, Powys, residence of D. Pugh, Chairman of the Oswestry & Newtown Railway.
Llanfair Grange	Monastic grange attached to Abbeydore. The same site as *Llanvair Grange*, see below and see text. (Fig. P85).
Llanfrechfa Grange	Two miles SE of Cwmbran, Gwent. Now a hospital.
Llangedwyn Hall	Eight miles WSW of Oswestry, Powys. Now demolished, it once was the property of *Sir Watkin Wynn* (which see).
Llanidloes	Name given by Contractor for the Llanidloes & Newtown Railway.
Llanrumney Hall	Midway between Newport and Cardiff, South Glamorgan. Now a restaurant.
Llanstephan Castle	Imposing ruins stand on a promontory above the Towy Estuary, just S of Llanstephan village, Dyfed.
Llanthony Abbey	Like *Malvern Abbey* (see text), this is more correctly Llanthony Priory. The ruined buildings are nine miles N of Abergavenny, Gwent.
Llantilio Castle	Two miles NW of Llantilio Crossenny, Gwent, and locally called White Castle. Once owned by G.W.R. Director Sir Henry Mather Jackson.
Llantwit (2)	Llantwit Fardre on the Llantrisant & Taff Vale Junction Railway.
Llanvair Grange	Private residence in the village of Bont, Gwent, six miles NE of Abergavenny. Same site as *Llanfair Grange* (above) and see text. (Fig. P84).
Llewelyn	Anlicised form of *Llywelyn*, which see.
Lleweni Hall	Three miles N of Denbigh, Clwyd.
Lloyd's	Named as a compliment to Lloyd's Register of Shipping. The unusual number of this engine, 100A1, applies normally to ships and means — (100) She is fit to put to sea — (A) She is built under Lloyd's rules — (1) Her anchors, cables, and mooring plans are approved by Lloyd's.
Llywelyn	Llywelyn ap Gruffydd, the last mediaeval Prince of Wales, slain at Builth in 1282 in battle against *King Edward I.*
Load Star	Older spelling of *Lode Star*, which see.

Lobelia	Handsome tall growing flower in blue, white and wine colour varieties.
Locke	Joseph Locke, Engineer of the Grand Junction and the London & Southampton Railways acted as a consultant to the G.W.R. Directors on several occasions.
Lockheed Hudson	U.S.A.-built aircraft used by R.A.F. in World War II.
Locust	Yet another quite unsitable name for a locomotive, see text.
Lode Star	The star that guides, usually known as Polaris.
London	The B.G. "Swindon" Class were named after places served by the G.W.R.
Longford Grange	One mile SW of Newport, Shropshire.
Longmoor	Name given on the Woolmer Instructional Military Railway, Longmoor, Hampshire.
Longworth Manor	Eight miles SW of Oxford.
Looe	Terminus of Liskeard & Looe Railway.
Lord Barrymore	G.W.R. Director 1904-25. 1st Baron Barrymore of County Cork, Ireland.
Lord Mildmay of Flete	G.W.R. Director, 1915-44, with country seat at Flete near Brent, South Devon.
Lord of the Isles	Well-known racehorse of the 1850's.
Lord Palmer	G.W.R. Director 1898-1906, Deputy Chairman 1906-43.
Lord Robartes (2)	Although he seems to have had no direct connection with the C.M.R., he had extensive mineral interests in the area served by the line.
Lord Tredegar	First Chairman of the A.D.R.
Lorna Doone	Character in a historical novel about N Devon by R. D. Blackmore.
Lotherton Hall	Eight miles E of Leeds, West Yorkshire. Owned by Leeds City Council.
Loughor	River which flows through the territory of the former Llanelly Railway.
Louisa	Alternative (or wrong) spelling of *Princess Louise*, one of Queen Victoria's daughters.
Lucan	Marcus Lucan, 39-65 A.D., Roman poet.
Lucifer	Alternative name for the *Morning Star.*
Lucretius	Roman poet, 99-55 B.C.
Ludford Hall	One mile SW of Ludlow, Shropshire. So described in "Pilgrimages to Old Houses" but known locally as Ludford House for many years.
Ludlow Castle	Well-preserved fortress which dominates this Shropshire town.
Lulworth Castle	Ten miles E of Weymouth, Dorset.
Luna	Roman equivalent of Greek mythological Goddess of the Moon, Selene.
Lupus	Latin for a wolf, one of six assorted names given to early Chester & Birkenhead locomotives for no obvious reason.
Lydcott Hall	Eight miles ENE of Barnstaple, North Devon.
Lydford Castle (3)	Stands on a spur overlooking the town of Lydford, seven miles NNW of Tavistock, Devonshire.
Lydham Manor	One mile NW of Bishop's Castle, Shropshire.
Lynx (2)	Fleet-of-foot animal evocative of the speed of a steam locomotive.
Lyonesse	In the Arthurian legends, it joined Lands End to the Scilly Isles but geologists cannot agree amongst themselves whether it ever existed.
Lyonshall Castle	Lyonshall, Hereford & Worcester, at junction of A44 and A480 roads.
Lysander	R.A.F. Reconnaisance aircraft in World War II.
Lyttelton	Port in New Zealand's South Island, visited on Royal Cruise, 1901, see text.
Lyttleton	Misspelling of the previous name.

Publications List

The Railway Correspondence and Travel Society

NEW RELEASE
THE LOCOMOTIVES OF THE GREAT WESTERN RAILWAY

Part 14

Names and their Origins
The Complete Preservation Story
Railmotor Services
War Service

The Great Western had a prolific naming policy with many subjects taken from outside its area of operations. This fascinating book recounts the stories behind the names, the reasons for and personalities behind particular choices, the subjects commemorated and personalities honoured by both the main Great Western Railway and its absorbed companies. 108 GW engines were 'called up' for war service. Their travels to and in the theatres of war are covered together with the five UK WD systems and depots involved. The steam branch line era is recalled with details of services operated by steam railmotors and auto trains, and the full preservation story is told. An amazing 157 Great Western locomotives have been preserved, of which 99 were rescued from Woodham Bros. yard at Barry. The engines themselves, and some of the trials and tribulations of preservation are covered together with some of the GW preservation sites and restored lines.

94 illustrations, 190 pages

This book completes the mammoth project which commenced before the birth of British Railways Standard designs more than twenty years ago. Many of the project team are no longer with us and we pay tribute to Ken Davies and the late Doug White for their major achievements in producing fourteen books that have become the accepted works of reference on their subject.

Part 13

Preservation and Supplementary Information

When this book was written in 1983 there were still 33 locomotives in Woodhams yard at Barry and many of the rescued engines awaited restoration. Interesting photographs of Barry and of preservation in progress bring alive the lack of facilities at preservation sites and gives a period flavour.

Laminated cover, 44 pages, 33 illustrations

Part 11

The Rail Motor and Internal Combustion Vehicles

Churchward was interested in the Dugald Drummond's experiments with steam railmotors in the southern counties and in 1903 the first two were built at Swindon. Eight constituent railways in addition to the Great Western itself built well over 100 units over the next decade. In 1906 40 electric multiple units were developed for underground services with the Hammersmith and City Railway, and in 1911 the first petrol railcar was experimented with. However it was not until 1933 that the first successful diesel railcar was developed. The principle was developed well by the GW, with express cars equipped with buffets and early examples of multiple coupling. At nationalisation, diesel shunters were under construction and a gas turbine locomotive was operating on a King express diagram. The story is presented in 25 pages with 39 illustrations.

Part 2

Broad Gauge

The Great Western was unique among the principal early railways of this country in having a gauge of 7ft 0¼in, adopted on Brunel's recommendations to provide faster and more comfortable travel than its narrow gauge contemporaries. Design constraints limited early progress and later benefits were clearly outweighed by the physical problems at the connecting points with other British railways built on the "narrow" 4ft 8½in gauge. This book presents the little known stories of the locomotives of the five broad gauge railway companies, including the 108 locomotives built latterly to later convertible designs. Readers familiar with the Barry saga will be taken back to its forerunner, the Swindon dump of 1892!
56 pages, 109 illustrations

The Locomotive History of the South Eastern Railway

The South Eastern Railway route to the Channel Tunnel and ports is currently in the spotlight with its potential capacity constraint on international traffic. This book gives the complete history of the railway and its engines from their origins to withdrawal as late as BR days, including the important Cudworth and Stirling designs.
226 pages, 104 illustrations

RCTS Order Form

* UK Post Free
Overseas add 20%

Trade Sales
6 Cherry Lane
Hampton Magna
WARWICK CV35 8SL

4/93

PLEASE SUPPLY

Title of Book		*ISBN No.*	**Price*	*No. Ordered*	*Amount*
Locomotives of the GWR:–					
Part 14	Names and their Origins, Railmotor Services, War Service, Preservation	0901115754	£10.95		
Part 2	Broad Gauge	0901115657	£4.95		
Part 11	Rail Motors	090111538X	£4.95		
Part 13	Supplementary Information	0901115606	£4.95		
Locomotive History of the South Eastern Railway		0901115487	£9.90		
Lord Carlisle's Railways		0901115436	£7.95		
Shildon–Newport in Retrospect		0901115673	£10.95		
LMS Locomotive Design and Construction		0901115711	£16.95		
Highland Railway Locomotives 1855–1895		0901115649	£12.95		
Highland Railway Locomotives 1895–1923		090111572X	£16.95		
Great Northern Locomotive History 1847–1866		0901115614	£12.95		
Great Northern Locomotive History 1867–1895		0901115746	£19.95		
Great Northern Locomotive History 1896–1911		090111569X	£19.95		
Great Northern Locomotive History 1911–1923		0901115703	£16.95		
Locomotives of the LNER:–					
Part 2B	Tender Engines Classes B1–B19	0901115738	£13.95		
Part 5	Tender Engines Classes J1–J37	0901115126	£9.95		
Part 6A	Tender Engines Classes J38–K5	0901115533	£9.95		
Part 6B	Tender Engines Classes O1–P2	0901115541	£10.95		
Part 6C	Tender Engines Classes Q1–Y10	090111555X	£10.95		
Part 7	Tank Engines Classes A5–H2	0901115134	£10.95		
Part 9A	Tank Engines Classes L1–L19	0901115401	£10.95		
Part 9B	Tank Engines Classes Q1–Z5	090111541X	£10.95		
Part 10A	Departmental Stock, Engine Sheds, Boiler and Tender Numbering	0901115657	£10.95		
Part 10B	Railcars and Electric Stock	0901115665	£13.95		
Easi–Binders for the RO – Wirex			£4.75		
Easi–Binders for the RO – Cordex			£4.75		
Society Tie – Red/Blue (state which)			£4.00		
LOW STOCK TITLES – ORDER NOW WHILE STOCKS LAST					
Locomotives of the GWR:–					
Part 1	Preliminary Survey	0901115177	£7.95		
Part 7	Dean's Larger Tender Engines		£7.95		
Locomotives of the London, Chatham & Dover Rly.		0901115479	£7.95		
Locomotive History of the SECR		0901115495	£7.95		
Locomotives of the LNER:–					
Part 3B	4–4–0 Tender Engines D1–D12	0901115460	£9.95		
Part 3C	4–4–0 Tender Engines D13–D24	0901115525	£9.95		

SIGNATURE TOTAL

LABEL:

Mac Iver	See *David MacIver.*
Madras	India was part of the British Empire when this locomotive was named.
Madresfield Court	Three miles ENE of Great Malvern, Hereford & Worcester.
Mafeking	Town in Cape Province, South Africa, besieged by the Boers 1899-1900.
Magi	Biblical Wise Men of the East.
Maglona	Name of the Roman Camp at Machynlleth.
Magnet	Title of a stagecoach service which ran between London & Weymouth and between London & Cheltenham.
Magpie	Not such an obvious choice for a C.&C. locomotive as *Heron*, which see.
Maid Marian	Character in the Robin Hood legends.
Maindy Hall	No Swindon record. Maindy is a suburb of Cardiff but no stately home or other building so named has been found.
Maine (2)	Ship purchased by the ladies of America and fitted out as a hospital ship to tend the wounded in the Boer War 1899-1902.
Majestic (2)	No. 3048 was almost certainly named after the battleship delivered a month before it was put to stock. The Birkenhead Railway name was given by the L.&N.W.R. after acquisition in 1860.
Malmesbury Abbey	Twelfth Century Abbey in the Wiltshire town of this name.
Malta	Mediterranean island visited on Royal Cruise, 1901, see text.
Malvern Abbey	The main Church in this Hereford & Worcester town is known as Great Malvern Priory but has been called an Abbey from at least 1721.
Mammoth	This extinct species of elephant typifies the power and size of a steam locomotive.
Manorbier Castle	This fine castle four miles W of Tenby, Dyfed, was one of the few Norman castles with spacious living accommodation. It was never attacked and later became the haunt of smugglers.
Manton Grange	One mile WSW of Marlborough, Wiltshire.
Marazion	Ancient Cornish town three miles ENE of Penzance.
Marble Hall	Probably the London pleasure gardens demolished about 1820 to make way for the Lambeth end of Vauxhall Bridge.
Marbury Hall	Three miles NNW of Whitchurch, Cheshire.
Marchioness	Wife of *Marquis* (of Blandford), which see.
Marco Polo	Venetian explorer who travelled through India and China and the Far East.
Margam Abbey	Ruined abbey three miles SE of Port Talbot, W Glamorgan.
Margaret (2)	Margaret Owen was the wife of John Owen, one of the promoters of the North Pembrokeshire & Fishguard Railway. No. 2478, as W.D. No. 172, received one of the temporary informal names given in World War II.
Marguerite	Member of the Crysanthemum family also known as the Ox-eye Daisy.
Marigold	Showy orange/yellow flower much favoured for borders.
Maristow	Country house of G.W.R. Director *Sir Massey Lopes*, at Roborough, Plymouth.
Maristowe	Misspelling of the previous name.
Marlas Grange	Monastic grange three miles N of Porthcawl, Mid Glamorgan, attached to *Margam Abbey.*
Marlborough	Wiltshire town from which the Duke took his name.
Marquis	Marquis of Blandford who rode on the locomotive of the first train over the Newtown & Machynlleth Railway.
Marquis Douro	Name given by L.&N.W.R.
Marrington Hall	Four miles E of Montgomery, Powys.
Mars	The two B.G. examples were named after the planet, repeated on the "Bulldog". The T.V. 0-6-0 took its name from the mythical God of War and the West Cornwall example could have been from either source.

Marwell Hall	Four miles NE of Southampton Airport, Hampshire. Now a zoo.
Mauritius	Island state in the Indian Ocean, visited on Royal Cruise, 1901, see text.
Mawddwy	Dinas Mawddwy was the terminus of the Mawddwy Railway.
Mawley Hall	One miles SE of Cleobury Mortimer, Shropshire.
Mazeppa (3)	The "Fire Fly" Class locomotive seems to have been named after the London to Hereford stagecoach (see text). The other two are more likely to have been from the 17th century Cossack chief brought to mind by a Lord Byron poem of 1819 and which, in turn, could have been the source of the stagecoach name.
Medea	Mythical daughter of the King of Colchis and priestess of the Goddess *Hecate*.
Medusa	Mythical *Gorgon* (which see), slain by *Perseus*.
Melbourne	Capital of the Australian State of Victoria. Visited on Royal Cruise, 1901, see text.
Melling	Almost certainly one of two people so named who had founded the Ince Forge, Wigan, in 1856 and had previously been at the Haigh Foundry.
Melmerby Hall	Four miles N of Ripon, N Yorkshire.
Melton Constable	Eight miles ENE of Fakenham, Norfolk; name given by purchasers of a former Cornwall Minerals Railway locomotive, see text.
Memnon	Mythological King of Ethiopa.
Mendip	Range of hills N of Wells, Somerset.
Mentor	Ithaca nobleman appointed by Odysseus as tutor to his son Telemachus.
Mercury (5)	In context, all five examples of this name refer to the planet.
Mere Hall	Four miles ESE of Droitwich, Hereford & Worcester.
Merevale Hall	One mile NW of Atherstone, Warwickshire.
Meridian	Pertaining to the sun at midday, a suitable name for one of the B.G. "Sun" Class.
Merion	Anglicised version of Meirion, first ruler of the County of Meirionydd (English Merioneth), see text.
Merkland	Believed to be a Scottish place name which a member of the Waddell family may have used for their home in the Llanelly area.
Merlin	Legendary magician at the Court of *King Arthur*.
Mermaid	An appropriate name for an 0-4-0ST used in the building of Fishguard Harbour.
Mersey (5)	This major river which flows into Liverpool Bay was an obvious choice for a locomotive on the parallel Birkenhead Railway; the other four examples also refer to the same river.
Merthyr (2)	Principal northern terminus of the Taff Vale Railway.
Meteor (3)	All three names referred to a "luminous heavenly body from outer space".
Metis	Mythical daughter of Oceanus and consort of Zeus.
Midas	Mythical son of Gordius; everything he touched turned to gold.
Middleton Hall	At Porthyrhyd, seven miles E of Carmarthen, Dyfed. The building was burnt down but there are plans to create a Welsh National Botanic Garden in the spectacular setting of this large site.
Miers	Richard Hanbury Miers, first Chairman of the N.&B.R.
Mignonette	Flowering shrub with sweet-scented blossoms.
Miles	John William Miles. G.W.R. Deputy Chairman, 1856-7.
Milford (2)	The Cambrian Railways' example would have taken its name from the Manchester & Milford Railway as it was delivered to its intended eastern terminus on the Llanidloes and Newtown Railway. The Pembroke & Tenby Railway terminated on the shore of Milford Haven.
Milligan Hall	Stood at Bishops Hull, one mile W of Taunton, until demolished in 1962.

Milo — The context suggests that this was the mythical Venus de Milo.
Minerva (2) — Mythical Roman Goddess of household arts.
Minnie — Unidentifiable name of a locomotive which may, or may not, have worked on the L.&M.M.R., see Part 10, page K225.
Minos — Mythical son of Zeus and owner of the Minotaur.
Misterton Hall — One mile SE of Lutterworth, Leicestershire.
Mobberley Hall — Five miles S of Altrincham, Cheshire.
Mogul — The Great Mogul once ruled over Hindustan.
Moloch — A particularly unpleasant mythical god whose worship was associated with human sacrifice. A strange choice indeed for a locomotive.
Monarch (2) — Rather than the obvious Royal connection, both locomotives to carry this name seem to refer to the London to Bristol stagecoach.
Monk — Name in keeping with *Druid*, an identical locomotive, when the original name *Birkenhead* was transferred to another locomotive.
Monmouth Castle — Ruins are to be found in this Gwent town's car park.
Montgomery — In the context of other names for this Class, could either be a Director's residence or a reference to the Earl of Montgomery.
Montgomery Castle — In the Powys town. Badly damaged after the Civil War in 1649.
Montreal — Canadian city at the junction of the Ottawa and St. Lawrence rivers; British Empire connection.
Monty (2) — The name given to "Dean Goods" No. 2399 (W.D. No. 94) and to No. 1956 after sale honoured Field-Marshal Viscount Montgomery of Alamein.
Moorsom — Engineer of the Birmingham & Gloucester Railway.
Morehampton Grange — Monastic grange attached to Abbeydore. Tantalisingly the map shows Morehampton Park Farm, Lower Grange Farm and Upper Grange, all within two mile of Abbeydore.
Moreton Hall — Four miles SW of Congleton, Cheshire, dating from 1841.
Morfa Grange — Monastic grange attached to Strata Florida Abbey.
Morlais Castle — Two miles N of Merthyr Tydfil, Mid Glamorgan.
Morning Star (3) — Poetic and popular name for the brightest planet, *Venus* but also sometimes used for the planet *Mercury*. While both are prominent in the eastern sky at or about dawn, somewhat illogically they both are also most prominent in the western sky at or about sunset.
Mortimer — See *Charles Mortimer.*
Moseley Hall — Four miles N of Wolverhampton where *King Charles II* once hid disguised as a wood-cutter. Known locally as Moseley Old Hall.
Mosquito — Yet another quite unsuitable name for any locomotive.
Mostyn Hall — Four miles NW of Holywell, Clwyd.
Mother Shuter — Nickname given by staff to No. 243 when working as the Worcester pilot.
Mottram Hall — Near Prestbury, Cheshire. Now a hotel.
Mountaineer (4) — An appropriate name for the B.P.&G.V. and N.&B. Fairlie locomotives which worked over steep gradients but out of place on a Cambrian 0-4-0ST. The R.R./T.V. example had been named by I. W. Boulton.
Mount Edgcumbe — Cornish family seat of the *Earls of Mount Edgcumbe.*
Mounts Bay (2) — Inlet on the S coast of Cornwall.
Mrs Jonson — Unofficial name given to a locomotive identical with *Ben Jonson.*
Mudlark — Name given by a Contractor.
Murdock — William Murdock, Scottish engineer, who first used gas for lighting.
Mursley Hall — Five milies WNW of Leighton Buzzard, Buckinghamshire.
Myrtle — Shrub with white flowers, blue-black berries and fragrant leaves.
Mytton Hall — Near Whalley in the Blackburn area of Lancashire.

Nanhoran Hall	Six miles SW of Pwllheli, Gwynedd. At first known as Plas Nant Horon, "anglicised" as Nanhoran Hall and now known as Nanhoron.
Nannerth Grange	Monastic grange on the banks of the Wye three miles NW of Rhayader, Powys, and attached to Strata Florida Abbey.
Nantclwyd	Name given by Contractor for the Denbigh, Ruthin & Corwen Railway, a place five miles S of Ruthin, Clwyd.
Napoleon	Napoleon III (1808-73). Ruled France from 1851.
Napoleon III	See *Napoleon*, above.
Narcissus	The daffodil, a bulbous rooted plant.
Natal	See *Natal Colony*, below.
Natal Colony	Province of S Africa, once part of the British Empire.
Neath	Southern terminus of the Neath & Brecon Railway.
Neath Abbey (2)	Partly restored extensive ruin of Cistercian Abbey one mile W of Neath, West Glamorgan, which gave its name to the 4-6-0. The T.V. example was more likely named after Neath Abbey Ironworks, where it was built. (Fig. P90).
Nelson (4)	The B.G. 0-6-0 was named after Horatio Nelson, victor in the Battle of Trafalgar, 1805; the 4-2-2's honoured R. R. Nelson, the G.W.R.'s Solicitor; the S.&W. name was given by I. W. Boulton.
Nemesis	Mythical Goddess daughter of Nyx, the personification of evil.
Neptune (2)	The 1838 locomotive was named after the planet; the B.G. 0-6-0 owed the origin of its name to Father Neptune, mythical God of the Ocean.
Nero	Claudius Nero, Roman Emperor 37-58 A.D.
Nestor (2)	A mythological wise old man. The West Cornwall Railway gave this name to an ex-L.&N.W.R. locomotive; the M.R.&C. example arrived from the M.S.&L.R. already bearing the name.
Newbridge (2)	19th century alternative name for Pontypridd, Mid Glamorgan.
Newlyn	Cornish fishing village one mile SSW of Penzance.
Newport (2)	Important junction between the G.W.R. South Wales main line and the "North & West" route and with several valley lines.
Newport Castle (2)	The ruins are visible from the main line east of Newport Station.
Newquay (3)	The northern terminus of the Cornwall Minerals Railway and one of the many West Country place names used for the "Duke" Class.
Newton	Newton-le-Willows, where the locomotive was built.
Newton Hall	No Swindon record. There are three buildings so named in Cheshire alone and others elsewhere.
New Zealand	Part of the British Empire when the locomotive was named.
Nightingale	The popular name of a summer bird with a particularly beautiful song.
Nimrod	Mythical great hunter.
Nipper (2)	Evocative name for a very small shunting locomotive. (See also *Scorcher*).
Nora Creina	19th century Irish beauty, subject of an air by her admirer, the poet Thomas Moore, and sung in most Victorian drawing rooms.
Norcliffe Hall	One mile SE of Manchester Airport, Cheshire.
North Aston Hall	Seven miles NW of Banbury, Oxfordshire.
Northiam	A station on the Kent & East Sussex Railway.
North Star (4)	Popular name for the star Polaris, carried by the four G.W.R. locomotives in turn from 1837 to 1957, a period of 120 years.
Northwick Hall	No Swindon record. There is one at the English end of the Severn Tunnel, one in Somerset and one in Hereford & Worcester.
Norton Hall	No Swindon record. There are several buildings with this name, both on and off the G.W. system.
Norwegian Monarch	Renamed to avoid confusion with "King" Class locomotives, see text.
Nunhold Grange	Private residence in Hatton ,Warwickshire.
Nunney Castle	Three miles SW of Frome, Somerset.

Oakley Grange	Belle Vue, Shrewsbury, Shropshire, now in use as flats.
Oakley Hall	Five miles WSW of Basingstoke, Hampshire.
Octavia	Of the three ladies in early Roman history so named, the context suggest the wife of *Nero,* which see.
Odney Manor	Although Swindon gave the location as Cookham, Berkshire, there never has been a building so named, or manorial lands. Lullebrook Manor (now a hotel) in Odney Lane, Cookham, presumably was the one in mind.
Ogmore Castle (4)	Two miles SW of Bridgend, South Glamorgan.
Oldlands Hall	Four miles SW of Crowborough, East Sussex.
Olton Hall	Four miles SE of Birmingham, West Midlands. Now the Olton Tavern.
Olympus	Mythical Greek mountain abode of the Gods.
Omdurman	In central Sudan, opposite *Khartoum* on the River Nile. Lord Kitchener defeated the Dervishes here in 1898.
One and All	Motto of the Duchy of Cornwall.
Ophir	Orient Line fast passenger ship chartered as a Royal Yacht, 1900, for Royal Cruise in 1901, see text.
Orion (4)	The mythological hunter. The "Bulldog" repeated the "Fire Fly" name.
Orleton	Home of R. C. Herbert, Cambrian Railways Director, at Wellington, Salop.
Orpheus	Thracian poet, said to be the son of the mythical *Apollo.*
Orson	The context does not help to decide if the name was taken from a Somerset farmer, subject of an 1809 novel, or from the 15th century French legend about a baby brought up by a bear.
Oscar	Oscar I, King of Norway 1844-59.
Osiris (3)	Mythical Egyptian God of the Underworld.
Ostrich (3)	Assumed evocative of the speed of a steam locomotive; perhaps the long neck was equated with the long chimneys common at the time!
Otho	(1815-67). First King of modern Greece, crowned in 1843 but deposed after a failed attack on Turkey.
Ottawa	Capital City of Canada; name given in the days of the British Empire.
Otterington Hall	Two miles S of Northallerton, N Yorkshire.
Overton Grange	Two miles S of Ludlow, Shropshire. Now a hotel.
Ovid	Roman poet, B.C. 43-17.
Owain Glyndŵr	Original Welsh version of (Owen) *Glendower*, which see.
Owen	William Owen, first Chairman of the Pembroke & Tenby Railway.
Owl	A nocturnal bird seems a strange choice for an 0-4-0WT.
Owsden Hall	Six miles ESE of Newmarket, Suffolk. Demolished in 1955.
Oxburgh Hall	Seven miles SW of Swaffham, Norfolk.
Oxford (2)	The broad gauge "Swindon" Class bore names of the principal places served by the G.W.R. The "Badminton" Class name may have referred to the University City.

Packwood Hall	Eleven miles SE of Birmingham, West Midlands.
Paddington	The name seems to refer to the station rather than the London district but there are several anomalies in the naming of Nos. 3443-52 (later 3381-90), as discussed under the names in question.
Pallas	A mythical *Titan*, the son of Crius.
Palmerston	British Prime Minister, 1855 and 1859-65.
Pandora (2)	The mythical first woman, reputedly created by Zeus to be troublesome!
Panthea	Relating to all the (mythical) Gods.
Park Hall	No Swindon record. May refer to Park Hall Hospital, Oswestry, served by a Halt but there are some other possibilities.

Parnassus	Mountain in Attica, Greece, where Zeus was worshipped.
Parwick Hall	Five miles N of Ashbourne, Derbyshire.
Pasha	Turkish Governor; possibly name taken from a ship built in 1846, see text.
Patshull Hall	Four miles WSW of Wolverhampton, West Midlands. The residence of G.W.R. Director *Earl of Dartmouth*, it usually was known as Patshull House.
Paviland Grange	Monastic grange on the Gower Peninsula, West Glamorgan, attached to *Neath Abbey.*
Peacock (2)	The B.G. locomotive took its name from Richard Peacock (1820-89) who partnered Charles Beyer in the firm of locomotive builders; No. 3450 was named after the bird with the beautiful patterned tail.
Pearl	The context does not help to decide if this was the name of a girl or that of the precious stone, but see also text.
Peatling Hall	Nine miles WNW of Market Harborough, Leicestershire.
Pegasus (3)	The mythical winged horse. The "Bulldog" repeated the "Fire Fly" name.
Pelican	With *Penguin*, rather different from the more elegant birds chosen as names for the other thirteen members of this batch of "Bulldogs".
Pelops	Mythical son of the Lydian King *Tantalus.*
Pembrey	The B.P.&G.V.R. passed through Pembrey, on the outskirts of Burry Port.
Pembroke (2)	No. 3386 (later 3412) was to have been named *Pembroke Castle* but it was realised that the oval cabside nameplates used on this batch would only take eleven letters and the name altered to *Pembroke*; by coincidence, the Earl of Pembroke had raised a regiment to fight in the Boer War so that the revised name fitted into the general context of the batch. In the P. & T. case, the name was that of the original western terminus until the line was extended to Pembroke Dock.
Pembroke Castle	Situate in the former County town, the castle has been extensively restored, having been sacked in the Civil War in 1648.
Penarth	Terminus of a Taff Vale branch.
Pendarves	Two miles S of Camborne, served by the West Cornwall Railway.
Pendeford Hall	Stood two miles from the centre of Wolverhampton, West Midlands. Demolished in 1952 and the site is now a caravan park.
Pendennis Castle (2)	On the western shore of the River Fal estuary, Cornwall. The last Royal Castle but one to surrender in the Civil War.
Pendragon	The chief King, by election, among the Ancient Britons.
Penguin	See comments on *Pelican*, above.
Penhydd Grange	Monastic Grange at Glyn-Neath on the West Glamorgan/Powys border, attached to *Margam Abbey.*
Penn	Appears to be John Penn (1805-78), whose firm specialised in ships' engines. President of the Institute of Mechanical Engineers 1858/9.
Penrhos Grange	Monastic grange at Raglan, Gwent, attached to Grace Dieu Abbey. The present building is in use as dog kennels.
Penrice Castle (3)	Ruined castle on Gower Peninsula, eleven miles SW of Swansea, which has a modern inhabited extension built on to the ruins.
Penwith	Name of the Duchy of Cornwall's ancient and historic Hundreds, situate at the western end of the West Cornwall Railway.
Penwyllt	Village almost at the summit of the Neath & Brecon Railway.
Penylan	Part of Cardiff through which the Contractor built a branch line of the Taff Vale Railway.
Penzance (3)	The western extremity of the West Cornwall Railway which became the end of the G.W.R. main line, accounting for all three names.

Peplow Hall	Ten miles NE of Shrewsbury, Shropshire.
Peri	Mythical malevolent elf or spirit.
Perseus (2)	Mythical son of the Greek God Zeus and of the Roman God *Jupiter*. The "Bulldog" repeated the "Iron Duke" name.
Perseverance (2)	Both names were given outside G.W. days and neither origin is obvious although the Directors of the financially-troubled Bishop's Castle Railway may have wanted to show that they were determined to carry on!
Pershore Plum	The local branch of the National Farmers Union invited the G.W.R. to name a locomotive to celebrate the centenary of the discovery of the plum growing wild in Pershore in 1827 and the Directors obliged.
Persimmon	Racehorse owned by the Prince of Wales, later *King Edward VII.*
Peterston Grange	Monastic grange at Peterstone Wentlooge, Gwent, six miles SW of Newport, attached to the Augustinian Canons of Bristol.
Petunia	Brilliant long-flowering plants with large showy trumpet flowers.
Peveril of the Peak	Title of Sir Walter Scott's novel based on the activities of Sir Geoffrey Peveril in relation to the Popish Plot of 1678.
Phlegethon (2)	One of the five mythical rivers of the Underworld.
Phoebus	Alternative name for the mythical God *Apollo.*
Phoenix (2)	From its context, the B.G. example was named after the mythical bird which arises from its own ashes. W.West & Sons, who worked the Cornwall Minerals Railway, also had an interest in the local Phoenix mine.
Pioneer (5)	Four of the locomotives with this name were related to new ventures, new designs, etc. (see text) but there is no obvious reason for its use in the middle of a batch of 102 B.G. 0-6-0's.
Pirate	See *The Pirate.*
Pisces	12th sign of the Zodiac. The Fish.
Pitchford Hall	Six miles SSE of Shrewsbury, Shropshire.
Plaish Hall	Six miles SW of Much Wenlock, Shropshire.
Planet	Having chosen *Mars* and *Mercury* for two of this batch of six locomotives, the makers seem to have run out of ideas.
Plasfynnon	Oswestry home of Thomas Savin's family.
Plaspower Hall	Once stood a mile N of Wrexham, Clwyd, known simply as Plas Power. "Plas" is a Welsh word for "Hall" and this bilingual mongrel actually would have translated as "Hallpower Hall"!
Plato	Greek philosopher, 429-347 B.C.
Plowden Hall	Four miles SE of Bishop's Castle, Shropshire.
Plutarch	Greek biographer of the 1st Century A.D.
Pluto (4)	Mythical King of the Underworld. The "Bulldog" repeated the "Fire Fly" name.
Plutus	Mythical son of Demeter and personification of wealth.
Plym	See *River Plym.*
Plymouth (3)	The "Bulldog" was one of a batch with predominantly West Country names; the two T.V. examples both referred to the Ironworks in Pentrebach, Merthyr, served by the T.V.R.
Plynlimmon	Wrong spelling of *Plynlimon*, see below.
Plynlimon	Mid Wales mountain east of Aberystwyth which is the source of both the River Severn and the River Wye.
Polar Star (2)	Popular name for the star Polaris.
Pollux (4)	All four were named after the mythical twin brother of *Castor*, which see.
Polyanthus	Flower which is a cross between a primrose and an oxlip.
Pontyberem	Intermediate station on the B.P.&G.V.R.
Pontypridd (2)	North western terminus of the Pontypridd, Caerphilly & Newport Railway.
Porth	Junction between the Rhondda and Rhondda Fach branches of the T.V.R.
Portishead (3)	Eastern terminus of the Weston, Clevedon & Portishead Railway.

Postlip Hall — Two miles SW of Winchcombe, Gloucestershire.
Poulton Grange — Five miles E of Cirencester, Gloucestershire.
Powderham — Estate of the *Earl of Devon*, four miles N of Dawlish Warren, South Devon.
Powderham Castle — Country seat of the *Earl of Devon* on the banks of the River Exe, eight miles SSE of Exeter.
Powerful (2) — Royal Navy cruiser which landed her guns to help relieve *Ladysmith* during the Boer War in 1900. (Fig. P86).
Powis Castle (3) — Fully preserved red sandstone castle on the outskirts of Welshpool, Powys, former seat of the Earl and Countess of Powis.
Precelly — Anglicised version of (Mynydd) Preseli, a mountain range in the area served by the railway.
Premier (2) — One of the first two locomotives to be delivered to the G.W.R. — on 25th November 1837 — had this most appropriate name, also used for the first locomotive to emerge from Swindon Works in February 1846.
President — President of France at the time the locomotive was built, see text.
Preston — First used by Contractor on construction of the L.B.&S.C.R.'s Cliftonville spur. Preston is a suburb of Brighton.
Preston Hall — No Swindon record. There are at least three in G.W. territory but the now British Legion Home near Canterbury, Kent, could have been chosen as it was once the home of Thomas Brassey.
Pretoria (2) — Capital of the Transvaal at the time of the Boer War.
Priam (2) — Mythical King of Troy at the time of the equally mythical Trojan War.
Primrose — Species of flower which includes also the primula and polyanthus.
Prince (2) — The first example could either have been the Prince Consort, or Queen Victoria's son *Prince Albert* who was two years old when this locomotive was built. The name of the 2-4-0ST of 1871 could have related to any one of the many Royal Princes alive at the time.
Prince Albert — Second son of *King George V*, later Duke of York and *King George VI.*
Prince Alfred — Son of Queen Victoria, later *Duke of Edinburgh.*
Prince Christian — Husband of Queen Victoria's daughter *Beatrice.*
Prince Eugene — Name given by L.&N.W.R. in post-Birkenhead Railway days.
Prince George — Son of *King George V*, later Duke of Kent. Killed on active service with the R.A.F. in 1942.
Prince Henry — Son of *King George V*, later Duke of Gloucester.
Prince John — Son of *King George V* who died in 1919 at the age of 13.
Prince of Wales (6) — The first three were named after the future *King Edward VII*, the V of R 2-6-2T after the future *King George V* and No. 4041 after the future *King Edward VIII.* When the old V. of R. name was put back on No. 9 in 1956 it was technically that of the future *King George V* although by then, Prince Charles had become Prince of Wales.
Princess — This S.W.M. name did not necessarily refer to any of Queen Victoria's many daughters but was "just a name".
Princess Alexandra — Daughter of Queen Victoria's son *Alfred* (1878-1942).
Princess Alice (2) — The Llanelli Railway locomotive took the name of Queen Victoria's daughter (1843-78). No. 4050 was named after the daughter of Queen Victoria's son Leopold (1898-1981).
Princess Augusta — Daughter of *Duke of Cambridge*, cousin to Queen Victoria (1822-1916).
Princess Beatrice (2) — Daughter of Queen Victoria, later Princess Henry of Battenberg (1857-1944).
Princess Charlotte — Daughter of Queen Victoria's daughter, the Princess Royal (1860-1928).

Princess Elizabeth	Daughter of Queen Victoria's daughter, *Alice* (1864-1918).
Princess Eugenie	Daughter of Queen Victoria's daughter, *Beatrice* (1887-1969).
Princess Helena (3)	Queen Victoria's daughter (1846-1923).
Princess Louise (2)	King Edward VII's daughter (1867-1931).
Princess Margaret	Daughter of Queen Victoria's son *Arthur* (1882-1920)
Princess Mary (2)	The Princess Royal, daughter of *King George V* (1897-1965).
Princess Maud	Daughter of *King Edward VII* (1869-1938).
Princess May (2)	The "family" name for the lady who became *Queen Mary*, wife of *King George V.*
Princess of Wales	Name given in 1896 to the (future) *Queen Alexandra.*
Princess Patricia	Daughter of Queen Victoria's son *Arthur* (1902-74).
Princess Royal (2)	Both names refer to Queen Victoria's eldest daughter Victoria (1840-1901).
Princess Sophia	Daughter of Queen Victoria's daughter, the Princess Royal (1870-1932).
Princess Victoria	Daughter of *King Edward VII* (1868-1935).
Priory Hall	In the town of Dudley, West Midlands. Now Council Offices.
Progress (2)	The N.&B. example was an appropriate name for the first Fairlie 0-4-4-0T. The S.H.T. example was named after sale.
Prometheus (5)	Mythical character who defended man against the God Zeus.
Proserpine	Mythical daughter of Zeus who was carried off by *Hades*, God of the Underworld.
Psyche	A beautiful heroine in Roman mythology.
Purley Hall	Four miles NW of Reading, Berkshire.
Pyle Hall	Once stood in North Cornelly, one mile SW of Pyle, Mid Glamorgan. Site now obliterated by M4 motorway.
Pyracmon	One of *Vulcan's* workmen in the mythical smithy on Mount Etna.
Pyrland Hall	Four miles N of Taunton, Somerset. Now a school.
Python	The name of a reptile seems inappropriate to an 0-6-0ST.
Quantock	Range of hills in Somerset, between Bridgwater and Watchet.
Quebec	Capital City of Quebec Province, Canada. Visited on Royal Cruise, 1901, see text.
Queen (6)	All six names were given during Queen Victoria's reign; the S.&B. example was named for working a Royal train in November 1852.
Queen Adelaide	Wife of *King William IV.*
Queen Alexandra	Wife of *King Edward VII.*
Queen Berengaria	Wife of *King Richard I.*
Queen Boadicea	Leader of the Iceni Tribe in East Anglia before the Roman Invasion.
Queen Charlotte	Wife of *King George III.*
Queen Elizabeth	Queen of England 1558-1603.
Queen Mary	Wife of *King George V* (1867-1953).
Queen Matilda	Wife of *King Stephen.*
Queen Philippa	Wife of *King Edward III.*
Queens Hall	Concert Hall in Portland Place, London, destroyed by enemy action, 1941.
Queen's Hall	The corrected spelling of *Queens Hall*, above.
Queensland	State in NE Australia.
Queen Victoria	Queen of Great Britain and the British Empire, 1837-1901.
Quentin Durward	Character in novel by Sir Walter Scott who joined the Scottish Archers to guard King Louis XI of France.
Quicksilver (2)	London to Devonport and London to Southampton stagecoach.
Racer	This appropriate name for an express passenger locomotive was enhanced when it was renamed *Glenside* (which see) when the horse won the Grand National in 1911.

Radcliffe	Name given by a Contractor and believed, by coincidence, to have worked at Radcliffe-on-Soar and at Radciffe, Manchester.
Raglan Castle	Red sandstone castle at Raglan, Gwent.
Ragley Hall	Magnificent stately home one mile S of Alcester, Warwickshire.
Ramsbury Manor	Two miles NW of Hungerford, near Marlborough, Wiltshire.
Ranger	Royal Officer in the Forest of Dean who arrested trespassers against the vert (vegetation) and venison and "received the profits of his office".
Ravelston	Waddell residence at Tumble, eight miles N of Llanelli, Dyfed.
Raven (2)	A batch of five South Devon 0-4-0ST's took the names of small British birds. The S.&W. name had been given by I. W. Boulton.
Raveningham Hall	Eleven miles S of Norwich, Norfolk.
Reading (2)	Both the B.G. 0-6-0 and the "Bulldog" related to the place and important junction station 36 miles west of Paddington.
Reading Abbey	Some ruins survive in Forbury Gardens, Reading.
Red Gauntlet (2)	Original version on the nameplates, which was altered to one word to correspond to the actual name of Sir Walter Scott's novel, but on No. 2983 only. The B.G. locomotive never was corrected.
Redgauntlet	A character in Scott's novel so named, based on the 1745 rebellion.
Redruth (2)	Original eastern terminus of the Hayle Railway.
Red Star (2)	Name sometimes given to *Mars* which more correctly is the "Red Planet".
Reepham	Twelve miles NW of Norwich, Norfolk. Named by purchaser of a C.M.R. locomotive.
Regulus	Marcus Regulus, Roman Consul 265-256 B.C.
Reindeer	The N.A.&H.R. chose the names of four fleet-of-foot animals for a batch of 2-4-0 passenger locomotives.
Remus (2)	Twin brother of *Romulus*, co-founders of Rome.
Rennie	Sir John Rennie, 1794-1874, a noted bridge builder.
Rescue	May refer to this small locomotive's work at Porthcawl Harbour, see text.
Resolute	Name given by Contractor.
Resolven Grange	Monastic grange five miles NE of Neath, West Glamorgan, attached to *Margam Abbey.*
Restormel	Site of *Restormel Castle,* which see.
Restormel Castle	12th century castle one mile N of Lostwithiel, Cornwall.
Rhea	Mythical Titaness daughter of Uranus.
Rheidol (2)	Dyfed river which rises on *Plynlimon* and joins the sea at Aberystwyth.
Rhiewport	Once the home of J. W. Jones (Director of the Aberystwyth & Welsh Coast Railway), near Welshpool, Powys.
Rhondda (4)	The river which parallels the T.V.R. from Pontypridd to Treherbert. Name also used by the B.G. 0-6-0. The A.D. example was named in anticipation of coal traffic expected to accrue from this valley.
Rhose Wood Hall	Built as a large bungalow adjoining Rhode Wood House, St. Brides' Hill, Saundersfoot, Dyfed, and known locally as Rhode Wood Hall. The buildings were combined to form Rhode Wood House Hotel.
Rhuddlan Castle	Red sandstone Castle two miles SSE of Rhyl, Clwyd.
Rhymney	As with *Gelly Gaer* (which see) the T.V.R. did not penetrate the valley of the River Rhymney.
Rignall Hall	Near Deddington, Oxfordshire.Now known as Rignell House.
Ringing Rock	English translation of Maenclochog, the original terminus of what became the North Pembrokeshire & Fishguard Railway.
Ripon Hall	Theological college at Oxford Univeristy.
Rising Star (2)	An early popular name for the planet *Venus*, the bright object in the eastern sky at sunrise.
River Fal	Cornish river entering the English Channel at Falmouth.

River Plym	Devon river entering the English Channel at Plymouth.
River Tamar	River which divides Devon and Cornwall and enters the English Channel SW of Plymouth.
River Tawe	River which rises in Powys and runs through West Glamorgan to enter the Bristol Channel at Swansea.
River Yealm	Devon river which enters the English Channel SE of Plymouth.
Roberts (2)	The B.G. locomotive took the name of a founder-partner of Sharp, Roberts & Co., locomotive builders of Manchester. No. 3387 (later 4133) honoured Field Marshal Lord Roberts, supreme British Commander in South Africa at the end of the Boer War in 1902.
Robertson (2)	No. 187 (later 2987) and 177 (2977) took their names from Sir Henry Beyer Robertson, G.W.R. Director 1905-45.
Robin Hood (3)	The S.&W. locomotive was named, for no obvious reason (see text) after the 13th century Sherwood Forest outlaw who stole from the rich to give to the poor. The two G.W.R. examples took the title of a novel by Sir Walter Scott.
Robins Bolitho	G.W.R. Director 1903-25 and a Cornish Banker.
Rob Roy (2)	A Walter Scott novel whose hero was a Scottish outlaw, Rob Roy MacGregor.
Rocket (2)	The B.G. "Sun" Class locomotive was one of a batch of eight where the other seven were named after lethal weapons. The S.D. example had *Comet* and *Meteor* as companions in its batch.
Rodwell Hall	Stood one mile SW of Weymouth but no trace of a building remains.
Roebuck	Reputedly the principal promoter of the Cornwall Minerals Railway.
Rolleston Hall	Two miles N of Burton-on-Trent, Staffordshire.
Romulus (2)	Twin brother of *Remus*, which see.
Rood Ashton Hall	Two miles SW of Trowbridge, Wiltshire. Once the residence of G.W.R. Director *Walter Long* but now a ruin.
Rook	One of five S.D. 0-4-0ST with names of small British birds.
Rosa	Could be a sister to *Ada* and *Una*, but see text.
Rosemary	Temporary informal name given in World War II.
Rose	The first four of the "1st Lot Metropolitan" bore the names of the floral emblems of Wales, England, Scotland and Ireland.
Rougemont (2) & Rougemont Castle	It is not clear why Rougemont was chosen for one of the "Iron Duke" Class or for No. 3022. Perhaps it refers to the site of Rougemont Castle (now a ruin) on high ground near Exeter Central Station but this is known as Northernhay Gardens.
Roumanian Monarch	Renamed to avoid confusion with "King" Class locomotives, see text.
Roundhill Grange	Almost five miles east of Castle Cary, Somerset.
Rover (3)	Origin unclear, possibly a reference to a locomotive which can rove far and wide. The second example actually gave its name to the celebrated "Rover" Class of B.G. 4-2-2's.
Royal (2)	See *Princess Royal.*
Royal Albert	Prince Albert, Queen Victoria's husband.
Royal Sovereign (3)	The name of No. 3050 would have referred to *Queen Victoria.* The other two uses were temporary names for special trains.
Royal Star (2)	Alternative name for the Star of Bethlehem.
Roydon Hall	Two miles NE of Tonbridge, Kent.
Ruby	The context does not help to decide if this was the name of a girl or that off the precious stone, but see also text.
Ruckley Grange	At Shifnal, Shropshire.
Rumney	Original name of the southern section of the Brecon & Merthyr Railway.
Runter Hall	Swindon records this as at Milcote, five miles SW of Stratford-upon-Avon but the only stately home in the area is Rumer Hall, so known since at least 1831.

Rushton Hall	Two miles NW of Tarporley, Cheshire.
Ruthin	Used on building of Denbigh, Ruthin & Corwen Railway.
Rydal Hall	One mile NNW of Ambleside, Cumbria. Now a conference centre.
Sabrina	Roman name for the River Severn.
Sagittarius	9th Sign of the Zodiac. The Archer.
Saighton Grange	Three miles SE of Chester. Now a college.
Saint Agatha	Virgin martyr of uncertain date and object of some fictitious Acts.
St. Agnes	Five miles NNE of Redruth, Cornwall.
Saint Ambrose	4th century Bishop of Milan, who composed and taught people to sing hymns.
Saint Andrew	Apostle and Patron Saint of Scotland.
St. Anthony	Four miles SSW of Falmouth, Cornwall.
St. Aubyn	Family name of the *Earl of Mount Edgcumbe.*
Saint Augustine	Probably refers to Augustine of Hippo (354-430), an intellectual and prolific writer with great influence on Christian thought.
St. Austell	Cornish town which took its name from Austol, a 6th century monk who founded the Church in Cornwall.
Saint Bartholomew	One of the twelve Apostles, about whom little is known.
Saint Benedict	Probably Abbot (c480-c550) and author of the Benedictine Rule, Patriarch of Western Monasticism.
Saint Benets Hall	See next item, nameplates at first omitted the apostrophe.
Saint Benet's Hall	About three miles SW of Bodmin, Cornwall, it was a lazar house in 1411 and more recently a private hotel, guest house and business premises.
Saint Bernard	Probably Cistercian Abbot (c.1090-1153) of Clairvaux, France, an order which founded over 50 religous houses in England and Wales.
Saint Brides Hall } *St. Brides Hall*	No Swindon record. Although the nameplates were "corrected" from "St." to "Saint", the only likely source is a property in the west of the County of Dyfed known as St. Brides Hall.
Saint Catherine	Katherine of Alexandria (c.4th century) was tortured on a wheel (hence Catherine Wheel). Patron Saint of a large number of our Churches.
Saint Cecilia	3rd century Roman martyr and patron saint of singers.
St. Columb	St. Columb Major and St. Columb Minor are both to the east of Newquay, Cornwall.
Saint Cuthbert	Bishop of Lindisfarne (c.634-87) whose name is recalled by Cubert, Cornwall.
Saint David	6th century Pembrokeshire bishop and Patron Saint of Wales.
St. Donats Castle	Three miles WSW of Llantwit Major, South Glamorgan. Now Atlantic College.
Saint Dunstan	Benedictine monk (909-88) who became Archbishop of Canterbury in 959.
Saint Edmund Hall	Part of Oxford University, where Edmund of Abingdon (c.1195-1240) was educated and later became Archbishop of Canterbury.
St. Erth	Four miles SW of St. Ives, Cornwall.
St. Fagans Castle	Four miles W of Cardiff, once the residence of the Earl of Plymouth.
Saint Gabriel	The Biblical Archangel.
Saint George	Palestinian soldier, martyr and Patron Saint of England (died c.303).
St. George (2)	Alternative version of last name, used on 2-2-2-/4-2-2 No. 3025.
St. Germans	Seven miles W of Plymouth on the main line to Penzance.
Saint Helena	(c.250-330) Mother of Constantine, the first Christian Roman Emperor.
St. Ives (2)	Resort and bay on the N coast of Cornwall.

St. Johns	Capital of Newfoundland visited on Royal Cruise, 1901, see text.
St. Just (2)	Either St. Just, seven miles W of Penzance or St. Just in Roseland, three miles NE of Falmouth, could have been the source of this name.
Saint Martin	Martin of Tours, to whom many English Churches are dedicated.
St. Mawes Castle	Fortress guarding the E entrance to Falmouth Harbour.
St. Michael (2)	Archangel said to have appeared to fishermen on St. Michael's Mount, Cornwall, in the 8th century, hence the dedication.
Saint Nicholas	4th century Bishop of Myra. The custom of giving children a present on his feast day led to the institution of Santa Claus.
Saint Patrick	(c.390-461) Bishop and Patron Saint of Ireland.
Saint Peter's Hall	Part of Oxford University.
Saint Sebastian	3rd century Saint of French birth but who was martyred in Rome.
Saint Stephen	First Christian martyr, stoned to death c.35 A.D.
Saint Vincent	Probably Vincent of Saragossa, martyred in Spain A.D. 304.
Salford Hall	Five miles NNE of Evesham, Hereford & Worcester.
Salisbury (2)	Lord Salisbury (1830-1903) was a leading Conservative politician when No. 1123 was built. The Mersey Railway named the A.D. locomotive.
Salopian	Either a native of Shrewsbury or the London to Shrewsbury stagecoach.
Salus	Mythical Roman Goddess of Health.
Sampson (2)	The Biblical strong man.
Samson (2)	From its context, the B.&M. example was an alternative spelling of *Sampson* (above). The G.W. example could be the same, or could be the large uninhabited island in the Scilly Isles.
Sandon Hall	Four miles N of Stafford.
Sappho	Poetess who lived about 600 B.C.
Sarum Castle	Once stood two miles NW of Salisbury, Wiltshire.
Saturn (2)	One of the four planets visible from the earth with the naked eye.
Saunders	See *Charles Saunders.*
Savernake	The context is no guide but could be the ancient Savernake Forest, Wiltshire, or a stately home in the Forest.
Scorcher	Name given by a later owner, but a 1' 11½" gauge 0-4-2T would hardly be capable of the turn of speed to justify such a title?
Scorpio	8th Sign of the Zodiac. The Scorpion.
Seagull	Popular name for medium to large, mainly white, marsh and sea birds.
Seaham	The husband of *Countess Vane* (which see) was Earl Vane of Seaham in the County of Durham and the eldest son of each Earl Vane has the hereditary title of Viscount Seaham.
Sebastopol (2)	Besieged in the Crimean War, 1855.
Sedgemoor	Marshy area W of Glastonbury, Somerset.
Sedley	Not traced either as a Director or Officer of the South Devon Railway but may refer to the main family in Thackeray's "Vanity Fair".
Sefton	Name given by the L.&N.W.R.
Seneca	Lucius Seneca, 4 B.C. - 65 A.D., Roman Stoic philosopher.
Severn (5)	All five uses of this name relate to the river which rises on Plynlimon and flows into the Bristol Channel. The T.V. name was chosen by the makers, see text.
Severn Bridge (2)	The name was first used by the Contractors who built the bridge and the S.&W. and S.B. Company named the second locomotive, delivered six months after completion of the bridge and its connecting lines.
Severus	(146-211 A.D.), Roman Emperor who mounted an expedition to Britain in A.D. 207, mainly known for strengthening Hadrian's Wall.

Seymour Clarke	Vice Chairman of the Banbury & Cheltenham Direct Railway and named by the Contractor.
Shah	Ruler of Persia.
Shakenhurst Hall	Eleven miles WSW of Kidderminster on the Shropshire/Worcester border.
Shakespeare	World famous Elizabethan playwright.
Shamrock	The Irish floral emblem, see note against *Rose* earlier.
Shannon (2)	The B.G. locomotive took its name from the Irish river, see text. The Wantage Tramway locomotive was named (on 17th June 1857!) after a contemporary steam frigate.
Sharp	Partner in the locomotive building firm of Sharp, Stewart & Co. Ltd.
Sharpness	Station on the Severn & Wye and Severn Bridge Railway.
Shelburne	G.W.R. Director 1858, Chairman 1859-63.
Shervington Hall	This very old house, seven miles from Ettington, Warwickshire, was at first known as Shennington Farm, then Shervington Hall and more recently Shennington House.
Shirburn Castle	One mile NE of Watlington, Oxfordshire.
Shirenewton Hall	Three miles N of Caldicot, Gwent.
Shooting Star	Popular name for a meteor, a small body which becomes incandescent on entering the earth's atmosphere.
Shotton Hall	No Swindon record. There is one example on Deeside and another in Shropshire.
Shrewsbury (2)	The B.G. "Swindon" Class bore names of the principal places served by the G.W.R. The "Badminton" Class 4-4-0 could have used the same source or could have been named after the school as this batch also included *Marlborough* and *Oxford.*
Shrewsbury Castle	Well preserved red sandstone castle on high ground near the station.
Shrugborough Hall	Alternative spelling of Shugborough Hall, five miles SE of Stafford and now the Staffordshire County Museum. Its exhibits include the wheels & motion of G.W. No. 252 (see also page P6).
Sibyl	Mythical prophetess who served *Apollo.*
Siddington Hall	Two miles SE of Cirencester, Gloucestershire.
Singapore	At southern tip of Malay Peninsula, visited on Royal Cruise, 1901, see text.
Sir Alexander	Sir Charles Alexander Wood, G.W.R. Director 1863-1890.
Sir Arthur Yorke	Chief Inspector of Railways 1900-1913. G.W.R Director 1914-1930.
Sir Daniel (2)	See *Sir Daniel Gooch.*
Sir Daniel Gooch	G.W.R. Locomotive Superintendent 1837-1864. Chairman 1865-1889.
Sir Edward Elgar (2)	Worcestershire composer of music, 1857-1934. See also text.
Sir Ernest Palmer (2)	See *Lord Palmer.*
Sir Felix Pole	G.W.R. General Manager 1921-1929. Resided at *Calcot Grange,* Reading.
Sir Francis Drake	Elizabethan sea-captain and national hero who led British ships against the Spanish Armada in 1588.
Sir George	Since no connection with the A.D.R. *Sir George Elliot* (See Part 10, page K10) has been found, the name is not relevant to this list.
Sir George Elliot	Director of the Alexandra (Newport & South Wales) Docks & Railway Company and promoter of the Pontypridd, Caerphilly & Newport Railway.
Sir Haydn	Sir Henry Haydn Jones, proprietor of the Talyllyn Railway prior to the advent of the Preservation Company.
Sirius	In context, mythical favourite dog of the hunter *Orion.*
Sir Ivor	Sir Ivor Guest, General Manager of Dowlais Ironworks which was served by the B.&M.'s Pant-Dowlais Central branch.
Sir James Milne	G.W.R. General Manager 1929-1947. Knighted 1932.
Sir John Llewelyn	G.W.R. Director 1892-1922.

Sir Lancelot — Knight of the Round Table in Arthurian legends.

Sir Massey — See *Sir Massey Lopes.*

Sir Massey Lopes — G.W.R. Director 1883-1908. His residence was *Maristow.*

Sir Nigel — See *Sir N. Kingscote.*

Sir N. Kingscote — G.W.R. Director 1895-1908. Fought at *Alma, Sebastopol* and *Balaclava* in the Crimean War 1854-1855.

Sir Redvers — Sir Redvers Buller, V.C., Supreme British Commander in the first years of the Boer War at the turn of the century.

Sir Richard Grenville — Elizabethan sea-captain who took on a whole Spanish fleet in H.M.S. *Revenge* in 1591.

Sir Robert Horne — G.W.R. Chairman 1934-1940. Later *Viscount Horne.*

Sir Stafford — Sir Stafford Northcote, Chancellor of the Exchequer, who was knighted in the year that No. 3368 (later 3356) was built.

Sir Walter Raleigh — Elizabethan sea captain said to have introduced potatoes and tobacco to this country.

Sir Watkin (3) — Sir Watkin Wynn, G.W.R. Director 1855-8, 1859-65, 1866-85 and also a Cambrian Railways Director, He lived at *Wynnstay*, which see.

Sir Watkin Wynn — Nephew of the above and a G.W.R. Director from 1885.

Sir W. H. Wills — See *Sir William Henry.*

Sir William Henry — G.W.R. Director 1888-1911. His full name Sir William Henry Wills was too long for a standard nameplate. Created (Lord) *Winterstoke* of Blagdon in 1906.

Sketty Hall — One miles WSW of Swansea and now used as offices.

Skylark — An abundant aerial songbird in Britain.

Slaughter — Partner in Stothert & Slaughter, locomotive builders, see text.

Smeaton (2) — Engineer who built Eddystone Lighthouse, off Plymouth, in 1756-9.

Snake — Name apparently chosen by the makers, Haigh Foundry, in company with *Viper.* It is difficult to imagine how these related to steam locomotives.

Snowdon — The highest mountain in Wales, visible from part of the Cambrian Railways system.

Sol — A mythological name personifying the sun.

Somerset — Family name of the Dukes of Beaufort, over whose Badminton Estate the G.W.R. were building the South Wales Direct Line when the locomotive was turned out.

Soughton Hall — Two miles N of Mold, Clwyd.

Spanish Monarch — Renamed to avoid confusion with "King" Class locomotives, see text.

Sparkford Hall — Four miles SW of Castle Cary, Wiltshire.

Speke Hall — 16th century National Trust property eight miles SE of Liverpool.

Sphinx — Mythical female monster, daughter of *Typhon*, with the body of a winged lion and the face of a young woman.

Spit Fire — Evocative name probably chosen by the makers of an early standard design. See also *Fire Ball*, etc.

Spitfire — R.A.F. fighter aircraft involved in the Battle of Britain, 1940.

Squirrel — David Davies bought this locomotive, so named, in 1858 for use on Cambrian Railways' construction contracts.

Stackpole Court — Once the stately home of *Earl Cawdor*, three miles S of Pembroke, Dyfed. Demolished and the land now owned by the National Trust.

Stag (4) — From the context, the G.W., S.D. and T.V. examples evoked the speed of the steam locomotive; the locomotive associated with the B.&M.'s early days was probably a small tank locomotive, which has defied attempts to identify it.

Stanford Court — Ten miles SW of Kidderminster, Hereford & Worcester.

Stanford Hall — No Swindon record. There are Halls so named NW of Lechlade, Gloucestershire, SW of Stratford-upon-Avon and at Lutterworth, Leicester.

Stanley Baldwin	Son of G.W.R. Chairman *Alfred Baldwin* and himself a Director 1908-17. Conservative Prime Minister 1923-4, 1924-9 and 1935-7.
Stanley Hall	Two miles N of Bridgnorth, Shropshire.
Stanway Hall	Stanway House, four miles SSW of Broadway, Hereford & Worcester, was known as *Stanway Hall* in its earlier years.
Starling	An abundant and familiar bird of countryside, town and garden.
Statius	Publius Statius, A.D. 45-96, Latin epic poet.
Stedham Hall	Two miles W of Midhurst, West Sussex.
Stella	Presumably a female relation of someone (known not to have been the Chairman) connected with the Pembroke & Tenby Railway.
Stentor	Mythical Trojan War warrior whose voice could be heard 50 miles away.
Stephanotis	Climbing plant with scented waxy flowers.
Stephenson (2)	The B.G. locomotive was named after George Stephenson (1781-1848), builder of the early locomotive *Rocket.* The name was given to the Port Talbot Railway locomotive by a later purchaser.
Steropes (2)	One of the mythological *Cyclops* (which see).
Stewart	Partner in the locomotive building firm of Sharp Stewart, of Atlas Works, Manchester, which adopted this title in 1852.
Stiletto	A short dagger with a slender blade. (A lethal weapon, see text).
Stokesay Castle	A fortified manor house at Craven Arms, Shropshire.
Storm King	Name evocative of the power and speed of a steam locomotive.
Stormy Petrel	The popular early British name for an oceanic sea-bird, the Storm Petrel but in this case more likely to be taken from the successful 1850's racehorse. Either way the name is evocative.
Stour	There were two rivers so named within G.W. borders, in Somerset and in Hereford & Worcester.
Stowe Grange	Monastic grange attached to Grace Dieu Abbey replaced by a private dwelling. One mile S of St. Briavels, Gloucestershire.
Stradey	Part of Llanelly although not near the route of the Llanelly Railway.
Stretcher	Informal name given by Powlesland & Mason men on account of its (reputed) long stovepipe chimney.
Stromboli (2)	Island volcano north of Sicily, no doubt evocative of a steam locomotive, especially at night.
Stuart	Family name of the Marquess of Bute, whose Cardiff West Dock and collieries were served by the T.V.R.
Sudeley Castle	One mile S of Winchcombe, Gloucestershire, it was once owned by Ethelred the Unready and remained Crown property until Tudor times, being Catherine Parr's palace. Now privately owned, it is in first class condition and is much visited.
Sultan (3)	Ruler of the Ottoman Empire but in these cases may be the name of a ship, see text.
Sun	Name probably given by the makers and by which this broad gauge class was known.
Sunbeam	As in the last case, name probably chosen by makers for one of the "Sun" Class.
Susan	Name given by the Bishwell Coal and Coke Co.
Swallow (3)	Used in the same batches as its French counterpart *Hirondelle*, which see.
Swallowfield Park	Five miles SSW of Reading, Berkshire, once the residence of *Charles Russell*, G.W.R. chairman 1839-55 and still owned by the family.
Swansea (2)	The "Bulldog" was named after one of the principal towns served by the South Wales expresses. The S.H.T. locomotive was named by a later purchaser.
Swansea Castle	Built in 1106, the western part of this castle has been repaired (the rest of the former site has been built upon) and can be seen in Swansea town (now city) centre.

Swedish Monarch	Renamed to avoid confusion with "King" Class locomotives, see text.
Sweeney Hall	Two miles SW of Oswestry, Shropshire.
Swift	Probably evocative of speed but the context is no guide. This bird, a common summer visitor, is the only species that breeds in Britain.
Swindon (3)	The B.G. 0-6-0 and the "Bulldog" were named after the place. No. 7037, the last of the "Castle" Class, was named as a tribute to both the works and the town, whose Borough Jubilee occurred in 1950.
Swithland Hall	Five miles S of Loughborough, Leicestershire.
Swordfish	Royal Navy bi-plane used in World War II.
Sydney	Capital of New South Wales, Australia, visited on Royal Cruise, 1901, see text.
Sylla	Misspelling of the mythical Scylla, an infamous sea monster who lured sailors to disaster.
Sylph	Mythical spirit of the air.
Taff (3)	All three were named after the river which rises in Powys and flows into the Bristol Channel at Cardiff.
Talbot	C. R. M. Talbot of Margam Park, Port Talbot, Chairman of the South Wales Railway 1849-63, G.W.R. Director 1863-90.
Talerddig (2)	Highest point of Cambrian Railways' system, between Newtown and Machynlleth. The second *Talerddig* was built for banking the incline.
Talgarth Hall	Stands at Talgarth, Powys.
Talisman (2)	The 4-6-0 took its name from Sir Walter Scott's novel in which *King Richard I* was cured by an amulet (Talisman). The Birkenhead locomotive was so named in its later L.&N.W.R. days.
Talybont	Name given by a previous owner, who worked the Talybont quarries.
Tamar (2)	See *River Tamar.*
Tangley Hall	Four miles N of Burford, Oxfordshire.
Tantalus	Mythical son of Zeus, punished in the Underworld by the sight of fruit out of reach above his head, origin of the word "tantalise".
Taplow Court	Early Tudor building, two miles E of Maidenhead, Berkshire.
Tarndune	Believed to be a Scottish place name and to have been used as the name of a house owned by the Waddell family at Tumble (on the L.&M.M.).
Tartar (3)	The context (see text) suggests either a ship built in 1840 or a racehorse, famous at the end of the 18th century rather than the conventional "member of a group of Central Asian peoples".
Tasmania	Part of the British Empire when the locomotive was built.
Tattoo	The makers' name for one of their standard designs, see text.
Taunton	Named after an important "cross roads" on the G.W. system.
Taunton Castle	Ancient fortress which now houses the Somerset County Museum.
Taurus (2)	The 2nd Sign of the Zodiac, the Bull, was clearly the origin of the B.G. name. The S.D. example is presumed to have come from the same source.
Tavy	This Devon river flows through Tavistock and into the Tamar Estuary.
Tawstock Court	Two miles S of Barnstaple, North Devon.
Tay	Scottish river which enters the sea at Dundee, see text.
Teifi	River in Dyfed which flows into Cardigan Bay.
Teign (2)	River in S Devon which flows into the English Channel at Newton Abbot.
Teilo	Welsh saint which gave his name to Llandeilo.

Telford	Thomas Telford, 1757-1834, bridge and canal builder.
Telica	From its context next to Vesuvius in the B.G. "Premier" Class it seems to be a Nicaraguan volcano but the choice is curious as it did not erupt from 1833 until the locomotive was built in 1846.
Tenby (2)	Eastern terminus of the Pembroke & Tenby Railway.
Tenby Castle (2)	Mediaeval castle which now houses this Dyfed town's museum.
Terrible	Royal Navy cruiser which landed her guns to help in the relief of *Ladysmith* during the Boer War in 1900.
Thalaba	Title of a poem by Robert Southey. Name given by the L.&N.W.R.
Thames (3)	River which rises in Gloucestershire and flows into the sea at Southend.
The Abbot	The Abbot of Hennaquhar was the title of a novel by Sir Walter Scott based on the days of Mary, Queen of Scots.
The . . . Monarch	See under the name of the country.
The Countess (2)	Wife of the Earl of Powis.
The Earl	*Earl of Powis*, which see.
The Earl of Dumfries	A hereditary title of the Marquess of Bute.
The Flying Flogger	Informal but evocative name of an early S.&C. locomotive.
The Gloucestershire Regiment 28th 61st	The County Regiment was formed from two old foot Regiments, the 28th and the 61st. The locomotive was renamed in April 1954 to commemorate the Regiment's role in the Korean War.
The Great Bear	A stellar constellation, chosen to match the name *North Star.*
The Lizard	Headland with famous lighthouse in South Devonshire.
Theocritus	Greek poet in 3rd century B.C.
The Pirate	Title of a novel by Sir Walter Scott.
The Queen	No. 3041 was so renamed in 1897, the year of Queen Victorias Diamond Jubilee. If the nameplates had read THE QUEEN! it could have been a permanent version of the Loyal Toast.
Theseus	A major figure in Greek mythology, best known for slaying the Minotaur, an equally legendary half-man, half bull.
The Somerset Light Infantry (Prince Albert's). *The South Wales Borderers*	Unlike The Gloucestershire Regiment 28th 61st (see above) there was no event to be commemorated in these two cases. Both regiments had their headquarters in Great Western territory and, while there is no official record, the approach may have come from the respective Commanding Officers; both locomotives carried the regimental crests.
The Wolf	Lighthouse on The Wolf rock between Lands End and the Scilly Isles.
Thirlestaine Hall	Near Cheltenham, Gloucestershire. Now building society offices.
Thistle	The Scottish floral emblem, see note against *Rose* earlier.
Thomas	Name given by a later purchaser.
Thornbridge Hall	Near Bakewell, Derbyshire. Now a conference centre.
Thornbury Castle (2)	Twelfth century castle nine miles N of Bristol.
Thornycroft Hall	At Chelford, near Macclesfield, Cheshire.
Throwley Hall	Six miles SE of Leek, Staffordshire. Ruined but to be rebuilt.
Thunderbolt	No. 3078 being next to No. 3079 *Shooting Star* implies an evocative name.
Thunderer (3)	This name, carried by two G.W. and one Birkenhead Railway locomotive also is evocative but the original G.W. *Thunderer* was a damp squib.
Tidmarsh Grange	One mile S of Pangbourne, Berkshire.
Tiger (2)	Both names were evocative of the locomotives' power and speed.
Timour	Mongol Chief (1335-1405) who conquered much of E Europe and Asia. His fame preceded the English translation of his life story in 1859 as the first locomotive to be so named was built in 1849.
Tintagel and Tintagel Castle }	Tintagel is a village in North Cornwall and its ancient castle stands on nearby Tintagel Head.

Tintern	Name given after the locomotive's T.V. days, by I. W. Boulton but probably with the village near Chepstow in mind.
Tintern Abbey	Extensive ruins of this Abbey stand four miles N of Chepstow, Gwent.
Tiny (2)	The South Devon example really was "tiny", see text. The B.&M. example had been named by Mr. Savin and it too was quite small.
Titan (2)	Member of a mythological tribe of strong men.
Titley Court	Nine miles W of Leominster, Hereford & Worcester.
Tityos	Mythological giant son of the Greek God Zeus.
Tiverton Castle	Red sandstone castle at Tiverton, mid-Devon, once owned by the *Earl of Devon.*
Tockenham Court	Nine miles WSW of Swindon, Wiltshire.
Toddington Grange	Seven miles S of Evesham, just inside the Gloucestershire border.
Tor	Welsh for "Rocky promontory". The highest point of the B.&M. system was at Torpantau, the "Tor above two valleys".
Torbay	Extensive inlet between Torquay and Berry Head, South Devon.
Tor Bay	The original spelling, as above, was altered to this correct version (as shown on the maps) in December 1903.
Tornado (4)	A suitably descriptive name for an express passenger locomotive.
Toronto	Name given when Canada was part of the British Empire. Capital of Ontario Province.
Torquay	The last of a batch of West Country names used on "Bulldog" 4-4-0's.
Torquay Manor	On high land overlooking Torquay. Presented by the U.S.A. to the Royal National Institute for the Blind after World War II. Always known locally as Torwood Manor.
Tortworth Court	Fourteen miles NNE of Bristol, near the village of Tortworth.
Totnes Castle	Mediaeval castle in this Devonshire market town.
Touchstone	Dark jasper, used for testing gold. An imaginative name for a 2-2-2 built for the Birkenhead Railway in 1840.
Towy (2)	The first Llanelly Railway locomotive so named was bought from the Contractor who built the Vale of Towy Railway. The second referred to the river which paralleled the Llandilo-Carmarthen line.
Towyn	Between Dovey Junction and Barmouth on the Cambrian Railways' coast line. The town's name has since reverted to the Welsh "Tywyn".
Toynbee Hall	Settlement in Commercial Street, E London. There never was a Hall as such, only St. Jude's Church Hall which once stood there.
Trafalgar (2)	Spanish cape off which *Nelson* won the Battle of Trafalgar in 1805.
Tranmere	Name given by the Mersey Railway.
Treago Castle	Small 15th century castle seven miles N of Monmouth, Gwent.
Treffrey	Misspelling of *Treffry*, see below.
Treffry	Squire J. T. Treffry, promoter of the Cornwall Minerals Railway.
Treflach	Branch line from Porthywaen built by Savin to serve his British Coal Pits at Coed-y-Glo, and not part of the Cambrian system.
Treforest (2)	Mining town on the T.V.R., S of Pontypridd.
Trefusis	Wooded area on the banks of the River Fal in Cornwall.
Tregeagle	In Cornish folklore, a Bluebeard who sold his soul to the devil.
Tregenna and Tregenna Castle }	The Castle, near St. Ives, N Cornwall, was built in 1774 and bought by the G.W.R. for use as a hotel in 1878. The name *Tregenna* was put on a "Duke" Class 4-4-0 in 1897 and *Tregenna Castle* on No. 5006 in 1927 but it was 1930 before the "overlap" was rectified (see note re page G17 on page P56).
Tregothnan	Country seat of Lord Falmouth, N of Falmouth, Cornwall.

Trelawny	Bishop, popular in Cornwall, made famous by a dispute with King James II in 1688 which led to him being imprisoned in the Tower of London.
Trellech Grange	Monastic grange attached to *Tintern Abbey*. The building now on the site, two miles W of Tintern, Gwent, is known by the anglicised name of Trelleck Grange.
Trematon Castle	In the same area as *Trematon Hall* (see below). A well-preserved 11th century castle owned by the *Duke of Cornwall* for the last 600 years.
Trematon Hall	Two miles W of Saltash Bridge in Cornwall.
Tremayne	Cornish family with extensive lands S of St. Austell.
Trentham Hall	Three miles S of Newcastle-under-Lyme, Staffordshire. Once the seat of the Dukes of Sutherland and demolished after they left in 1911.
Tre Pol and Pen	In the Cornish language, Tre is a town, Pol a pool and Pen a headland.
Tresco	See *Isle of Tresco.*
Tresco Abbey	Monastery on the *Isle of Tresco* in the Scilly Isles.
Tretower Castle (2)	Adjoins Tretower Court, just north of the junction between the A40 and A479 roads, in Powys.
Trevithick (2)	Richard Trevithick, 1771-1833, a pioneer in the development of the steam locomotive. The B.G. locomotive was named in this context but the "Duke" name would have referred to him as a Cornish mining engineer.
Trevor Hall	One mile WSW of Ruabon, Clwyd.
Trinidad	When the locomotive was named this West Indian island was part of the British Empire.
Trinity Hall	No Swindon record. It could either refer to a part of Cambridge University or a Hall so named at Newmarket, Suffolk.
Trio	Name given by Thomas Brassey, Locomotive Contractor to N.A.&H.R.
Trojan	An evocative name for the apparent strength of a small shunting locomotive.
Troy	Temporary informal name given in World War II for no obvious reason, see text.
Truro	Eastern terminus of West Cornwall Railway.
Tubal Cain	Its place next to *Hercules* in the names of the Cambrian "Small Goods" Class suggests the strength of the Biblical Tubal Cain as a blacksmith.
Tudor Grange	Now a school and recreation centre in Solihull, West Midlands.
Tweed	River which enters the North Sea at Berwick-on-Tweed, see text.
Twineham Court	Ten miles NNW of Brighton, West Sussex. Demolished and now farmland.
Tylney Hall	Seven miles NE of Basingstoke, Hampshire.
Tyne	River which enters the North Sea E of Newcastle-on-Tyne, see text.
Typhon	Mythical monster with a hundred heads resembling burning snakes.
Ulysses (2)	The Roman equivalent of the mythical Greek Odysseus, one of the leading personages in Homer's "Iliad".
Umberslade Hall	Three miles SW of Hockley Heath, Birmingham, West Midlands. Sometimes referred to locally as simply "Umberslade".
Una	Could be a sister to *Ada* and *Rosa*, but see text.
Underley Hall	At Kirkby Lonsdale, Cumbria. Now a school.
Upton Castle	The original castle, at Pembroke, Dyfed, has been incorporated in a later building.
Usk (3)	The B.&M./Cambrian locomotive was named after the river which enters the Bristol Channel at Newport, Gwent. The A.D. example was named by a previous owner and could be the river

	or the Gwent town.
Usk Castle	Ivy-covered residence at Usk, Gwent.
Uxbridge	Locomotive used by the contractor who worked goods traffic on the C.M.&D.P. until the Company's own locomotives arrived.
Vancouver	When the locomotive was named, Canada was part of the British Empire.
Vanguard	No. 98 (later 2998) was the first true Churchward standard locomotive and it seems odd that the name, apparently evocative of progress, was only given after it had run for four years.
Velindre	A location five miles NE of Kidwelly from which the Gwendraeth Valleys Railway hoped to obtain traffic in quarried stone but which did not materialise.
Venus (3)	Planet visible with the naked eye in the morning and evening.
Vesper	Alternative name for the planet *Venus*, which see.
Vesta	Mythical Roman Goddess of the Hearth.
Vesuvius	Volcano near Naples, here evocative of locomotive steam and smoke.
Victor (2)	Probably carried by the locomotive when acquired from a Contractor as the British Royal Family, the source of all contemporary Llanelly Railway names, did not include anyone called Victor. The Severn & Wye locomotive was so named at Bryn Navigation Colliery.
Victor Emanuel	King of Sardinia from 1849 and of Italy too, 1861-1878.
Victoria (8)	This name was used eight times, six of which definitely relate to Queen Victoria. The A.D.R. locomotive was named by the Mersey Railway. For the B.P.&G.V. example, see text.
Victoria and Albert	Queen Victoria and her husband, Prince Albert.
Victory	Appropriate name for a locomotive built shortly after the end of World War I.
Violet (2)	The broad gauge example was a shy violet among a batch mainly named after much larger flowers and shrubs. The Birkenhead locomotive was named by the L.&N.W.R.
Viper	An unsuitable locomotive name, see comments against *Snake*, above.
Virgil	Publicus Virgil, 70-19 B.C., a Roman poet.
Virginia	Temporary informal name given in World War II, see text.
Virgo	6th Sign of the Zodiac. The Virgin.
Viscount Churchill (3)	G.W.R. Chairman, 1908-34.
Viscount Horne	See *Sir Robert Horne.*
Viscount Portal	G.W.R. Director 1927-40 and 1944/5. Chairman 1945-7.
Viso	Alpine peak. Name given by M.S.&L.R.
Vixen	A female fox sits oddly among 32 Classical/Mythological names given to these B.G. 0-6-0's.
Voiara	Temporary informal name given in World War II, see text.
Volante	Probably the name of a racehorse, see text.
Volcano (2)	Name evocative of the fire, smoke and steam of a locomotive.
Voltigeur (2)	French name for light infantryman but in these contexts, more likely a famous mid-19th century racehorse, epitomising locomotive speed.
Volunteer	This name almost certainly commemorates the part played by Cambrian railwaymen in the Crimean War and who were members of the Third Montgomeryshire (Railway) Rifles.
Vulcan (6)	The B.G. example took its name from its Vulcan Foundry birthplace and the other five were all the mythical Blacksmith to the Gods.
Vulture	Presumably chosen as it epitomised the speed of a locomotive.

Wales (2)	The original name *Prince of Wales* was shortened by the Llanelly Railway.
Wallsworth Hall	Two miles N of Gloucester.
Walter Long	G.W.R. Director 1893-1905. Lived at *Rood Ashton Hall.*
Walter Robinson	G.W.R. Director 1869-95, Deputy Chairman 1895-1910.
Walton Grange	One mile SSE of Much Wenlock, Shropshire.
Walton Hall	No Swindon record. May be an ancient residence near Stratford-upon-Avon.
Walton Park	One mile NNW of Clevedon, Somerset.
Wantage Hall	Part of Reading University, Reading, Berkshire.
Wardley Hall	Five miles S of Bewdley, Hereford & Worcester.
Wardour Castle	In Wiltshire, four miles ENE of Shaftesbury (Dorset). A ruin in the care of English Heritage, usually known as Old Wardour Castle.
Warfield Hall	Two miles N of Bracknell, Berkshire.
Warhawk	Appears to be a man of action, see *Warrior* below.
Warwick Castle	Very fine and well-preserved castle at Warwick, Warwickshire which had been the home of the *Earls of Warwick* for centuries.
Warlock (3)	The dictionary definition "Sorcerer" does not fit the context. It may have been a ship built in 1840, see text.
Warrior	The first four of the B.G. "9th Lot Goods", *Warrior, Warhawk, Gladiator* and *Lagoon* (vice Laocoon) had names with a combative theme.
Wasp	This completed a set of totally unsuitable names for locomotives.
Watcombe Hall	Two miles N of Torquay, S Devon, now called Brunel Manor. A house started here by Brunel in 1847 was not completed and the present house, built in 1870, probably used the same foundations.
Waterford	Marquis of Waterford, whose second daughter married the Duke of Beaufort, over whose land G.W.R. track was being laid at that time.
Watt	See *James Watt.*
Wavell	1st Earl Wavell, 1883-1950, commanded British Forces in Italy, 1944.
Waverley (2)	Jacobean gentleman in a novel of this name by Sir Walter Scott.
Wear	River which rises in the Pennines and enters the North Sea at Sunderland.
Weasel	An apt name for a very small 0-4-0WT.
Weaver	Presumably the Lancashire River Weaver although some distance from any point on the Birkenhead Railway.
Wellington (4)	The "Castle" Class locomotive was named after the R.A.F. bomber used in the early days of World War II. The other three took their names from the Duke of Wellington, the "Iron Duke".
Werfa	The Werfa Dare Colliery in the Dare Valley was served by the T.V.R.
Western Star	One of the popular names for the planet *Venus*, see also *Morning Star.*
Westminster Abbey	The Collegiate Church of St. Peter, Westminster.
Westminster Hall	Part of the Houses of Parliament, originally built in 1097.
Westol Hall	Eight miles E of Worcester in Hereford & Worcester.
Weston (2)	Western terminus of the W.C.&P.R.
Weston-super-Mare	The name could have been given to promote this seaside resort.
Westward Ho	Almost certainly refers to a brand of tobacco made by Sir. W. H. Wills' Bristol firm. All the 4,200 men who worked on the Gauge Conversion in May 1892 were given a 2oz. pouch of this tobacco and the locomotive was an early form of mobile advertisement for the product.
Westwood Hall	No Swindon record. There was a Westwood Park at Droitwich, Westwood House at Leek, Staffordshire and a large estate in

	Wiltshire.
Weymouth	Most of this batch of "Bulldogs" had names with a West Country flavour.
Whaddon Hall	Six miles E of Buckingham and known locally as Whaddon Park.
Whetham	Charles W. Whetham, Director of the Milford Railway.
Whitbourne Hall	An 1860 building in Greek Revival style, eight miles WNW of Worcester.
White	Lt. General Sir Gordon White V.C., Commander-in-Chief of British Forces at the beginning of the Boer War, 1898.
White Horse	Presumed to be a reference to the ancient figures cut in chalky hillsides like the White Horse visible from the main railway line at Westbury, Wiltshire.
Whitmore	Name given by a later owner.
Whittington	Cambrian Railways station two miles E of Oswestry.
Whittington Castle	Fortress in the Shropshire village near Oswestry.
Whixall	Whixall Moss (or bog) lies E of Bettisfield and was a formidable obstacle encountered in building the Cambrian's Oswestry, Ellesmere & Whitchurch Railway.
Whorlton Hall	Three miles ESE of Barnard Castle, County Durham.
Wick Hall	A building dating from 1700, one mile S of Radley, Oxfordshire.
Wightwick Hall	Three miles W of Wolverhampton and known locally as Wightwick Manor.
Wigmore Castle (2)	At Kingsland, Hereford & Worcester. Once owned by the ancestors of *Charles Mortimer*, a long-serving G.W.R. Director.
Wild Fire	Evocative name probably chosen by the makers of an early standard design. See also *Fire Ball*, etc.
Wilkinson	Sir Joseph Wilkinson, G.W.R. General Manager, 1896-1903.
Willesley Hall	Near Ashby-de-la-Zouch, Leicester. Demolished 1953 and grounds now a golf course.
Willey Hall	Three miles E of Much Wenlock, Shropshire.
William	Name given by Manchester Collieries Ltd.
William Dean	Locomotive & Carriage Superintendent at Swindon, 1877-1902. (See Fig. M6).
Willington Hall	Two miles NW of Tarporley, Cheshire.
Will Scarlet	Character in the Robin Hood legends.
Will Shakspere	Probably an alternative early spelling of *Shakespeare* (which see) or even a misspelling. (A contemporary account gives his son's name as Shakspeare).
Wimpole Hall	Eight miles SW of Cambridge. Now a National Trust property.
Winchester Castle	Ruined castle on a hill at the W end of the City of Winchester.
Windsor (2)	Including the B.G. locomotive with thirteen of the principal places served by the G.W.R probably was a compliment to its Royal Family connection. The S.H.T. name was given by a later owner.
Windsor Castle (3)	Magnificent Royal Palace in the centre of the Berkshire town.
Winnipeg	When the locomotive was named, Canada was part of the British Empire. The capital of the province of Manitoba.
Winslow Hall	Six miles SE of Buckingham.
Winterstoke (2)	When Sir William Henry Wills was elevated to the peerage in 1906, he took the title of Lord Winterstoke. He was a G.W.R. Director 1888-1911.
Wirral	The Chester & Birkenhead Railway ran for most of its length through the Wirral Peninsula.
Witch	A strange choice of companion for *Prince, Queen* and *Sylph*!
Witchingham Hall	Elizabethan mansion near Reepham, Norfolk, which once had a moat.
Witherslack Hall	Four miles ENE of Grange-over-Sands, Cumbria.
Wizard	An out-of-context name for a B.G. express passenger locomotive.
Wolf (2)	Fleet-of-foot animal evocative of the speed of a locomotive.

Wolf Hall	Tudor style house one mile ENE of Burbage, Wiltshire.
Wollaton Hall	16th century house, now the Natural History Museum in Nottingham.
Wolseley	Viscount Wolseley was Commander in Chief of the British forces 1895-1901, which included the period of the Boer War and was Adjutant-General in 1884 at the time of the Sudanese War.
Wolseley Hall	At Rugeley, Staffordshire. Home of the *Wolseley* family for seven centuries until demolition in 1965.
Wolverhampton (2)	The B.G. "Swindon" Class bore names of the principal places served by the G.W.R., used in the same context on the "Bulldog".
Wood	Associate of George Stephenson and judge at the Rainhill trials.
Woodcock Hall	At Dolphinholme, Lancaster.
Woollas Hall	Four miles S of Pershore, Hereford & Worcester.
Woolston Grange	At Winchcombe, Gloucestershire, and known locally as Woolstone Grange.
Wootton Hall	No Swindon record. There are several buildings so named in G.W. territory and elsewhere.
Worcester (2)	No. 158 bore the name of the Capital City of Worcestershire for three months in 1895, whereupon it was transferred to No. 3027.
Worsley Hall	In Manchester, home of the Earls of Ellesmere since the Norman conquest.
Wraysbury Hall	Two miles SW of Staines, Middlesex. Originally named "Rivernook".
Wrekin (2)	Shropshire landmark, a name used both by the S.&B. and the S.&C.
Wrottesley Hall	Four miles W of Wolverhampton, West Midlands.
Wycliffe Hall	Part of Oxford University.
Wye (5)	River which rises on Plynlimon and flows into the Severn at Chepstow, Gwent.
Wyke Hall	Four miles NW of Shaftesbury, Dorset.
Wyncliffe	Original name of the Fishguard Bay Hotel.
Wynnstay (2)	Stately home near Ruabon, Clwyd, principal residence of the Watkin Wynn family. Now used as Lindisfarne College.
Xerxes	King of Persia who lived from B.C. 519 to 465.
Yataghan	Single-edged cut and thrust sword from Turkey and the Balkan States. (For comment on such lethal weapons, see text).
Yiewsley Grange	Situate in High Street, Yiewsley, Middlesex. Now belongs to a Light Engineering firm.
Zebra (2)	The other four members of this batch of the "Sun" Class were faster on their feet but evocation of speed must have been intended. The South Devon managed to mix it up with five other names of a very different character.
Zeno	Ancient Greek philosopher.
Zetes	One of the mythical twin sons of Boreas, the God of the Wind.
Zillah	Biblical character, wife of Lameck and mother of *Tubal Cain*. Name used by the Birkenhead Railway in both 1840 and 1857.
Zina	A Persian ruler who lived about 2500 B.C.
Zopyrus	Follower of the Persian Darius at the time of the attack on Babylon.
Zygia	Surname of the mythical Goddess Hera, presiding over marriage. The name is appropriately derived from the verb meaning "to yoke".

AMENDMENTS TO THE INDEX IN PART 12 AS AMENDED BY PART 13

The notes on Page N39 and the table at the foot of page N44 of Part 13 are included here so that all amendments to the "Index of Names", Tables B and C on pages M154-M177 of Part 12 appear below:-

TABLE B: Add *Conqueror* of the "Atbara" Class, with page reference P57.

TABLE C: Delete reference to *Conqueror* on page M159 as this name was not carried in service.

The page reference for *Druid*, page M161, is B24, not B21.

The name *Dusty* on page M161 was given in error for *Duty*, see below.

The page reference for *Saint Peter's Hall* on page M173 is M96, not MXX.

Also on page M173, *Sallopian* should read *Salopian* and *Saus* should read *Salus*.

The name *Bobs* was (fortunately) omitted from page M157 as it has since been found that the name never applied, as explained on page P50.

The following additions to Table C should be made in alphabetical order on the pages shown. Only the names *Beacon* and *E. H. Llewellyn* refer to the period of M.&S.W.J. and G.W.R. ownership and all the others should be inset on the style of *Acton* on page M155.

Insert on page				
M155	*Alexander*	"2301"	N40	(D69)
M156	*Ashburton*	Collett 0-4-2T	N8	(F22)
"	*Beacon*	M.&S.W.J.	P107	—
M157	*Betty*	"2301"	N40	(D69)
"	*Bulliver*	Collett 0-4-2T	N8	(F22)
M158	*Canada*	C.M.&D.P.	P111	—
"	*Cardiff*	Cardiff	P112	(K88)
"	*Casabianca*	"2301"	P112	(D69)
"	*Casablanca*	"2301"	N40	(D69)
"	*Churchill*	W.C.&P.	P113	(K267)
"	*City of Birmingham*	"2301"	N40	(D69)
M160	*Daisy*	"2301"	P115	(D69)
M161	*Duty*	B.P.R.&P.	P117	(C81)
"	*Earl of Berkeley*	"Earl"	N8	(G17)
M162	*E. H. Llewellyn*	"Bulldog"	P119	(G19)
"	*Fagan*	"2301"	P120	(D69)
"	*Flockton Flier*	"6400"	N11	(E84)
"	*Flying Fortress*	"2301"	N40	(D69)
"	*Francis*	"2301"	N40	(D69)
M163	*Gert*	"2301"	P122	(D69)
"	*Goliath*	"4200"	N10	(J38)
M164	*Hecla*	Cambrian	P125	—
"	*Hero*	Cambrian	P125	—
M165	*Hove*	Llanelly	M71	(C48)
"	*Jean Ann*	"2301"	N40	(D69)
"	*Jean Barbara*	"2301"	N40	(D69)
M168	*Margaret*	"2301"	N40	(D69)
"	*Moloch*	B.&E.	P135	(B36)
M169	*Monty*	"2301"	N40	(D69)
"	*Nipper*	F.&B.	P136	(C78)
M172	*Rosemary*	"2301"	N40	(D69)
M173	*Scorcher*	F.&B.	P145	(C78)
M175	*Thomas*	S.W.M.	P150	(K255)
M176	*Troy*	"2301"	P152	(D69)
"	*Uxbridge*	C.M.&D.P.	P153	—
"	*Virginia*	"2301"	N40	(D69)
"	*Voiara*	"2301"	N40	(D69)
"	*Wavell*	"2301"	N40	(D69)
M177	*Whitmore*	Cambrian	P155	(K56)

ENDPIECE: A DIRGE OVER THE BROAD GAUGE

Gone is the Broad Gauge of our youth — its splendid course is run —
It has fought the battle nobly, but the narrow gauge has won;
Alas! for good Sir Daniel — Alas! for bold Brunel —
They are resting from their labours — they are sleeping — it is well!
Yes! 'tis well that they had left us, and had landed on that shore
Where the battle of the gauges could vex their souls no more.
They loved their own Great Western, 'twas their glory and their pride,
The Queen of all our railways, with gauge so fair and wide.

How grand their lordly engines! we knew and liked them well —
The mighty *Eupatoria* — the graceful *Hirondelle*
There was dashing *Nora Creina* — and *Lightning* dazzling bright —
The fiery-breathing *Dragon* — the *Swallow* swift of flight —
There was *Hebe* fair and saucy — and who will e'er forget
The *Emperor* right royal, and the lively *Estafette*?
Resistless ran *Tornado* — and *Amazon* could breast,
The fury of the tempest in her journeys to the west:
Lord of the Isles — *Great Britain* — these and a thousand more
That sped along the railway in the good old days of yore:
We have lost them! we have lost them! but we loved them in their prime,
And their names shall long re-echo through the corridors of time.

But evil days were coming on, and clouds were gathering fast
That threatened soon to darken all the glories of the past.
Men trembled, scarce believing, but whisperings were heard
Of *narrow* gauge — we shuddered at that ill-omened word.
Oh never — never may it be — we all exclaimed at first,
But louder grew the whisperings — too soon we knew the worst.
From Paddington the fiat came — 'The old broad gauge MUST go' —
From north to south, from east to west, men reeled beneath the blow.

Directors are omnipotent — with them no man can cope —
Our utmost consolation was that while there's life there's hope.
And so it lingered year by year — at length they fixed the day —
One Friday, in this year of grace, in the merry (?) month of May:
Oh! that the western counties such day should ever see!
Such night on railways ne'er has been, and ne'er again shall be.
The last — last train would leave Penzance when clocks that night struck ten;
No broad gauge train should afterwards be seen by western men.

The hour peeled forth upon our ears a melancholy knell —
'Twas over now — the train moved off mid grief that none could quell.
O sense of desolation that fell on all that crowd,
Where some were weeping quietly, and others sobbed aloud.
The last hand-shake was given — the last farewell was said —
All felt that now their fond broad gauge was numbered with the dead.

The sad, though calm, Inspector waved mournfully his hand,
As he bravely faced his journey drear throughout the western land;
And as each stationmaster saw the red lamps, burning bright,
Recede into the darkness, and silence of the night,
He wrung his hands in agony, and smote upon his breast,
To see that last of broad gauge trains pass to its final rest.

by A. B. Berry, of Penzance, May 1892.

Great Northern Locomotive History

Volume 1 1847 to 1866

The Great Northern Railway was formed in 1847 and, with locomotive engineers such as Sturrock, Stirling, Ivatt and Gresley, has always been in the forefront of locomotive interest. The most important previous locomotive history was that of G. F. Bird published in 1910 and now very rare. It suffered from the difficulty in obtaining access to official records and has both omissions and errors. The present book is the result of ten years work by Norman Groves consulting such records and can be considered a definitive work on the subject. It covers the period from 1847 to 1866 and is followed by three more books bringing the story up to 1923.

Board covers, 124 illustrations, 123 pages.

Volume 2 1867 to 1895 — The Stirling Era

In 1938 the Society sponsored the first chartered rail tour. Stirling Single No.1 hauled this train and is featured in this book, together with the other 710 engines designed by Patrick Stirling during his 28 years as Locomotive Engineer of the GNR. Stirling's locomotives were among the most handsome ever built, including four- and six-coupled tank and tender designs in addition to his Singles classes. The complete history of all variations of design, livery, allocation and use of each engine is included.

Board covers, 237 illustrations, 311 pages.

Volume 3A 1896 to 1911 — The Ivatt Era

Henry Ivatt was faced with a major challenge when he succeeded Patrick Stirling as Locomotive Engineer of the Great Northern Railway in 1896. Much increased train weights had led to late running expresses and expensive double-heading of both these and mineral workings. London suburban trains needed more capacity.
With no available completed designs, the sensational and little known purchase of American 2-6-0s was first adopted whilst Ivatt developed his plans. He designed a fleet of fourteen classes totalling 767 locomotives over a period of fifteen years which revolutionised the east coast system's operations. Most memorable was undoubtedly the first British 4-4-2 Atlantic designs.
Author Norman Groves gives the complete history of each design, its modifications and the locomotives' allocation and work.

Board covers, 202 illustrations, 235 pages.

Volume 3B 1911 to 1923 — The Gresley Era

Gresley's priority on appointment was for larger locomotives to handle the growing mineral, goods and commuter traffic. His first design was a Churchward inspired Mogul, the H2, followed by the acclaimed 01 2-8-0. After the First World War, he developed three-cylinder versions of these and the superb and fondly remembered N2 0-6-2Ts. The culmination of Gresley's GNR work were his A1 Pacifics, that brought a revolution on the ECML that was to last fifteen years. The complete history of all 328 engines built during the Gresley era are fully described in this book — which completes the series — including sections on GNR engine order numbers, Doncaster works numbers, engine headlamp codes, shed codes and complete statistics of all GNR engines.

Board covers, 67 illustrations, 134 pages.

RCTS Book List

Order from:
Hon Assistant Publications Officer
Hazelhurst
Tiverton Road
Bampton
Devon EX16 9LJ

*UK Post Free
Overseas add 20%

Please Quote Ref:PUBS 4 **4/93**

Title of Book	ISBN No.	*Price
Locomotives of the GWR:-		
Part 14 Names and their Origins, Railmotor Services, War Service, Preservation	**0901115754**	**£10-95**
Part 2 Broad Gauge	0901115657	£ 4-95
Part 11 Rail Motors	090111538X	£ 4-95
Part 13 Supplementary Information	0901115606	£ 4-95
Locomotive History of the South Eastern Rly	0901115487	£ 9-90
Lord Carlisle's Railways	0901115436	£ 7-95
Shildon-Newport in Retrospect	0901115673	£10-95
LMS Locomotive Design and Construction	0901115711	£16-95
Highland Railway Locomotives 1855-1895	0901115649	£12-95
Highland Railway Locomotives 1895-1923	090111572X	£16-95
Great Northern Locomotive History 1847-1866	0901115614	£12-95
Great Northern Locomotive History 1867-1895	0901115746	£19-95
Great Northern Locomotive History 1896-1911	090111569X	£19-95
Great Northern Locomotive History 1911-1923	0901115703	£16-95
Locomotives of the LNER:-		
Part 2B Tender Engines Classes B1-B19	0901115738	£13-95
Part 5 Tender Engines Classes J1-J37	0901115126	£ 9-95
Part 6A Tender Engines Classes J38-K5	0901115533	£ 9-95
Part 6B Tender Engines Classes O1-P2	0901115541	£10-95
Part 6C Tender Engines Classes Q1-Y10	090111555X	£10-95
Part 7 Tank Engines Classes A5-H2	0901115134	£10-95
Part 9A Tank Engines Classes L1-L19	0901115401	£10-95
Part 9B Tank Engines Classes Q1-Z5	090111541X	£10-95
Part 10A Departmental Stock, Engine Sheds, Boiler and Tender Numbering	0901115657	£10-95
Part 10B Railcars and Electric Stock	0901115665	£13-95

LOW STOCK TITLES - ORDER NOW WHILE STOCKS LAST

Title of Book	ISBN No.	*Price
Locomotives of the GWR:-		
Part 1 Preliminary Survey	0901115177	£ 7-95
Part 7 Dean's Larger Tender Engines		£ 7-95
Locomotives of the London, Chatham & Dover Rly	0901115479	£ 7-95
Locomotive History of the SECR	0901115495	£ 7-95
Locomotives of the LNER:-		
Part 3B 4-4-0 Tender Engines D1-D12	0901115460	£ 9-95
Part 3C 4-4-0 Tender Engines D13-D24	0901115525	£ 9-95